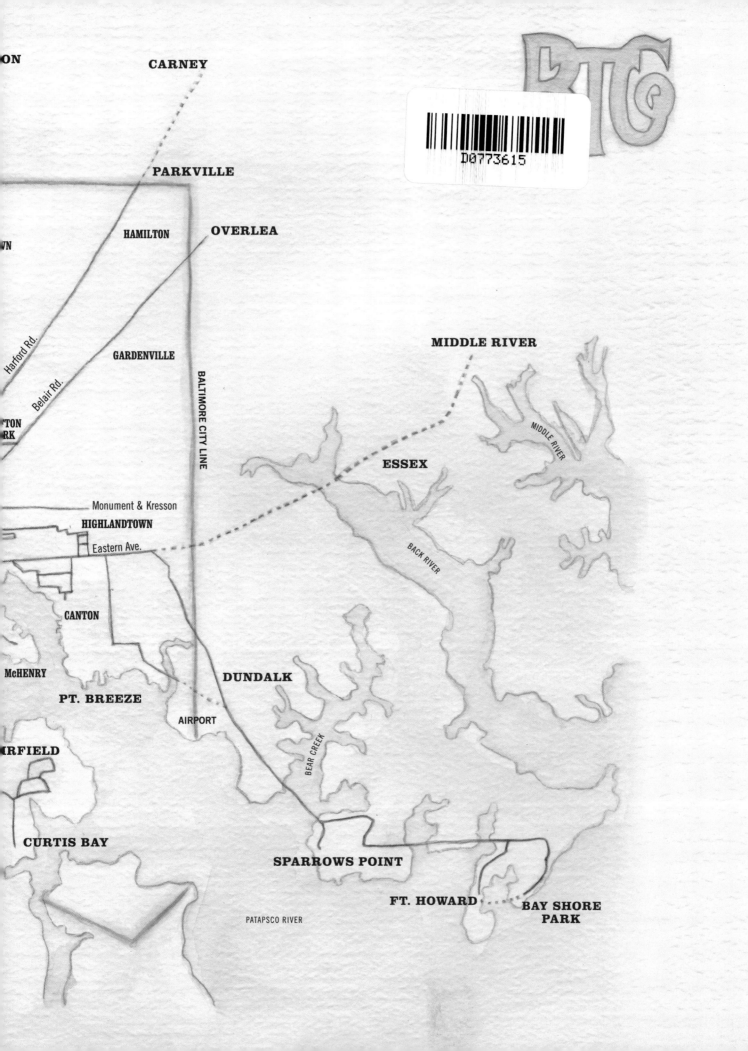

BALTIMORE

THE POSTWAR YEARS

STREETCARS

PUBLISHING FOR THE WORLD
125 Years
THE JOHNS HOPKINS UNIVERSITY PRESS

BALTIMORE
THE POSTWAR YEARS
STREETCARS

╬ **HERBERT H. HARWOOD, JR.** ╬

WITH A NEW FOREWORD BY

PAUL W. WIRTZ, TRUSTEE,

BALTIMORE STREETCAR MUSEUM

THE JOHNS HOPKINS UNIVERSITY PRESS

BALTIMORE AND LONDON

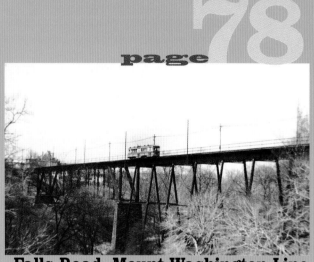

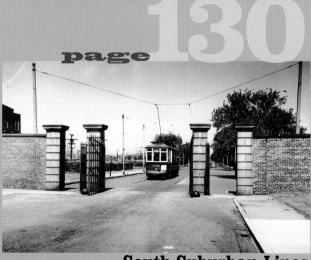

contents

A canopy of trolley wires overhung Fayette and Holliday Streets in September 1946. Beneath those wires ran the rails of a unique streetcar system. Seen here is one of Baltimore's unusual two-section articulateds picking its way across the switchwork

Foreword

For millennia, even large cities were small enough and so configured that people could live close to their workplaces, stores, and other destinations, with no need for transportation other than their legs. Carriages and sedan chairs were for the indolent wealthy. Then, with the coming of the Industrial Revolution, living close to your work became less possible. Thus, the early nineteenth century saw the invention of the omnibus, a half-wagon, half-coach that carried passengers on established routes within cities—the real beginning of urban public transportation.

But technological change was quick, and even before the first omnibuses began jouncing over rough city streets, the new railroads had proved that metal rails not only provided a far smoother ride but also reduced friction, allowing horses to haul heavier loads faster. The country's first urban street railway appeared in New York in 1832, but it took more than two decades for the idea to catch on broadly. Baltimore was one of the earlier innovators; it opened its first horsecar line in 1859, and streetcar mass transit had arrived.

Thanks in large part to the reasonably reliable horsecar service—hampered only by heavy snows and, in one case, by a citywide epizootic that sickened most of the horses for several weeks—the city grew quickly. The 1880 census showed that 39 percent of the state's population now lived in Baltimore. And the small, slow horsecars were now proving unequal to the demands of the growing community. Efforts were made to devise some type of replacement for animal power, and in 1885 Baltimore pioneered the nation's first electrically powered streetcar service—a line from present Twenty-fifth Street to Hampden. Unfortunately, the new service suffered from technical inadequacies and expired in a few years. In 1888, however, an innovator and inventor named Frank J. Sprague developed a practical electric streetcar and built the first such system in Richmond, Virginia. Sprague's system used an overhead wire and so-called trolley pole to collect current, with electric motors mounted on the car's trucks. It instantly became the standard, and cities quickly scrambled to adopt the new technology.

Baltimore was not so sure. The City Council at first was reluctant to grant franchises for lines using such a dangerous medium as electricity, and the city had a brief and expensive flirtation with cable cars—a purely mechanical system using understreet cables to move the cars at a more or less constant speed. It was only a few years, however, before the council was convinced of the value of electric cars; franchises were granted and new lines proliferated. As they did, they were often promoted by separate companies that would sometimes compete with one another on nearby streets. Over a short period of years, though, these merged with one another. Finally, by the beginning of the new century, all were brought together in a single street railway company, the United Railways & Electric Company (UR&E).

The UR&E quickly proved itself an aggressive and progressive company and soon became one of the best customers of Philadelphia's J. G. Brill Company, the nation's largest streetcar builder. By 1918 almost nine hundred large, double-truck, wood-bodied Brill semiconvertible cars journeyed the city streets, and they became integral parts of the city's life for more than three decades. Although some modern cars arrived in the 1930s and 1940s, the old Brills seemed to characterize Baltimore almost to the end.

During the first part of the twentieth century, the streetcar was the greatest single influence on Baltimore's physical growth, as it was for most major cities. This growth took place along what originally had been turnpikes leading from the city but were now streetcar routes through the city and into its new suburbs. No longer was it necessary to live within a mile or two of one's workplace. Almost two thousand streetcars, operating on more than four hundred miles of lines, brought everyone to within a half-hour or less of the city's center. During rush hours some routes scheduled cars at one-minute intervals. Off-peak service, even on lightly traveled lines, had intervals of not more than fifteen minutes.

And for decades, the streetcars were everyone's transportation: bankers and clerks, factory workers and managers, domestics, schoolchildren, letter carriers, firemen, and policemen all rode together (the last two groups rode free while in uniform). There was no racial segregation on the streetcars, in a city where residences and schools were segregated. In short, the streetcars were a truly democratic institution—unlike today, when all too often many people equate public transportation with the poor and elderly.

All was not well, however. By the 1930s, despite its investment in some of the newest and finest equipment, the streetcar company began steadily losing its riders to private autos. Only World War II, with its gas rationing and swarms of war workers needing to get to the factories and yards, prevented the disappearance of the streetcar. Providentially, the Baltimore Transit Company had not scrapped cars as declining ridership and new equipment had made them surplus during the late 1930s. Many of the same wooden streetcars that had served in World War I were available to soldier on, transporting hordes of workers—tens of thousands to the Bethlehem-Fairfield shipyards, builder of more Liberty ships than any other U.S. yard; more tens of thousands to the Bethlehem Steel mills and shipyard at Sparrows Point; and many thousands more to other industries in the area.

But when the war ended, the automobiles again multiplied and the inevitable decline in passengers resumed and accelerated. As the number of autos increased, there appeared a new profession—the traffic engineer. The traffic engineer saw his mission as one of moving large volumes of motor vehicles more rapidly, not reducing their number and not encouraging streetcars.

Baltimore's first full-time professional traffic engineer said that "the only trouble with streetcars is that they are on the street." As an aid to expediting the flow of autos, two-way streets were made one-way, making them incompatible with existing streetcar trackage. The replacement of streetcars with buses—touted as "free wheel" vehicles, less likely to impede traffic—was pushed by newspapers, politicians, car drivers, and, not least, bus manufacturers. Furthermore, as new suburbs were created, farther and farther from the city, they outstripped the existing car lines; the cost of extending the lines was prohibitive for a company already losing money. The cheaper bus, although slower and less reliable than a rail vehicle, provided a quick fix for all these problems.

Wholesale bus conversions followed, and in November 1963 the last streetcar went out of service. People no longer traveled within the urban area on steel wheels. As autos continued to proliferate and the move to the suburbs accelerated, traffic increased and inevitably worsened despite expressways and the best efforts of traffic engineers. Twenty years later the first increment of a part-subway Metro opened; eleven years after that, the first segment of a new light rail line appeared. The two lines, now complete (but with extensions contemplated), approximate about one-seventh of the streetcar mileage that was abandoned. Today some people in Baltimore ask why the streetcars were ever done away with.

In this book Herbert Harwood, a noted rail historian and photographer, has brought to us the Baltimore streetcar system in the autumn of its life, a survivor of the past but still an important transportation asset. This book should fascinate not only Baltimoreans but also anyone interested in a mid-twentieth-century American city with most of what had been one of the nation's larger and better-run street railway systems still largely intact.

—Paul W. Wirtz
Trustee, Baltimore Streetcar
Museum

Preface and Acknowledgments

Virtually every large city had its streetcar era, each one usually with its own distinct personality. But Baltimore's seemed to be a summary of them all. Whatever features other trolley systems had, Baltimore usually had them somewhere—plus a few oddities of its own. Perhaps that is not surprising, since Baltimore always had an unsettled personality. It was partly Old South, partly nineteenth-century New England mill town, partly midwestern industrial city, partly cosmopolitan international port, partly office and regional financial center, and partly its own peculiar blend of big city and small town atmospheres.

Whatever the reason, Baltimore's urban transportation system in the first half of this century had diversity and charm almost unequaled elsewhere. Its trolleys lurched through a labyrinth of complex downtown trackwork and along single-track rural jerkwater lines; they threaded through steel mills and shipyards; they rolled past woods, inlets, ponds, amusement park roller coasters, and—of course—endless row houses built from the 1700s to the 1930s. There were the latest PCC streamliners and numerous turn-of-the-century wood-bodied Brill semiconvertibles, which were among the oldest streetcars operating anywhere. (The PCC was named for its designer, the Electric Railway President's Conference Committee.)

And, as a basically conservative, rather dowdy city—sixty years ago, at least—Baltimore seemed to cling to a streetcar system that in many ways was already archaic. Indeed, it took two "foreign" elements to shake the system into the mid–twentieth century—a traffic commissioner imported from Denver and transfer of control of the transit company to Chicago. The changes that followed were undoubtedly inevitable, but much was lost in the process.

This book's purpose is to communicate the diversity and atmosphere of Baltimore's streetcar era as it was in its best-remembered days and at the same time to show something of what the "streetcar city" itself looked like. Thus, this is neither a history nor a dissertation on operations and equipment. For general history, there is Michael R. Farrell's *Who Made All Our Streetcars Go?* (originally published in 1973 by the Baltimore Chapter of the National Railway Historical Society), which was revised and reprinted by Greenberg Publishing Company in 1992 as *The History of Baltimore Streetcars.* More recently, Father Kevin Mueller covered the era of the Baltimore Transit Company, from 1935 onward, in his privately published *The Best Way to Go: The History of the Baltimore Transit Company* (1997). For those interested in the details of the cars themselves, two complementary books cover the full range: *Early Electric Cars of Baltimore,* by Harold E. Cox, and *Baltimore Streetcars, 1905-1963,* by Bernard J. Sachs, George F. Nixon, and Harold E. Cox. Some of these may still be purchased at the Baltimore Streetcar Museum, which also offers some specialized publications dealing with streetcar routes and structures.

In compiling this book, the author quickly discovered that, in the 1940s and early 1950s, the enthusiast-photographer with a "scenic sense" was a rare bird. Happily, he found several who not only had that sense but also were generous and helpful. Most notable were Edward S. Miller of Pittston, Pennsylvania, and George J. Voith of Baltimore. Without them, documentation of Baltimore's streetcar era would be far poorer. Equally helpful were Robert S. Crockett, Robert W. Janssen, Fred W. Schneider III, Robert M. Vogel, and J. William Vigrass. David B. Ditman read the messy manuscripts and corrected errors, both factual and grammatical.

Special thanks are also owed to Paul W. Wirtz, who helped nurse this second edition into print by critically reviewing the original 1984 edition and suggesting corrections and additions.

My own debt to them all is extreme; more important, anyone interested in Baltimore's history owes them much.

BALTIMORE
THE POSTWAR YEARS
STREETCARS

Gutted by the Great Fire of 1904, downtown Baltimore was quickly rebuilt and then changed little over the next five decades. Its streetcars seemed equally ageless. This scene looking west on Baltimore Street at the intersection of Saint Paul and Light Streets dates to the mid-1930s; except for the style of the autos, the scene could be any time from World War I to after World War II. No. 5541 is typical of the vast fleet of so-called semiconverties built by Philadelphia's J. G. Brill Company between 1905 and 1919. Here it is working the No. 16 line to Fells Point; behind is the city's traditional "hub," Baltimore and Charles Streets, marked appropriately enough by The Hub department store and the Baltimore & Ohio Railroad's general office building. R. V. MEHLENBECK

Depending on how you define it, Baltimore's streetcar era lasted between 73 and 104 years before the last car ran in 1963. But Baltimoreans who remember and rode the cars usually think of them as they were in the 1940s, the last decade that they dominated the city's public transportation. This book illustrates that period and looks not only at the trolleys themselves but also at the world in which they operated—obviously a very different world from that of today. Before looking at the pictures, however, we should stop to take a snapshot of the city itself and its streetcar system as it was in one of those last "typical" years—say, 1945.

This was the Indian summer of the Baltimore streetcar. World War II had brought a surge of riders, many of whom soon would be driving cars again. A large fleet of fine, streamlined PCC cars recently had been delivered and, despite some Depression-era line abandonments, the system still was large, varied, and busy. But winter came quickly. By the late 1950s, only two car lines would be left, battered remnants hanging on mostly because the company could not afford to buy buses to replace them. In early November 1963, these, too, finally died, and their tracks quickly vanished under asphalt. Today it takes some looking to find physical traces of the transportation system that built, nourished, and expanded the city. Memories hold some images of the cars and lines, but these, too, are increasingly indistinct.

So let us go back briefly to the way things were in 1945, at the end of the war and the beginning of the end for the streetcar. Looking at Baltimore today, it is often impossible even to picture the environment in which the streetcars ran, much less the system itself. In many ways the Baltimore of 1945 was a wholly different city: different physically, socially, philosophically, and economically. The most obvious difference was in its outward appearance, especially downtown and in the outlying suburbs. There was nothing resembling an express highway—no Beltway, no Interstate 95 or Jones Falls Expressway, no Baltimore-Washington Parkway, no harbor tunnels. The autos and trucks, including all north-south and east-west intercity traffic, fought their way through city streets. And those streets were mostly two-way, often occupied by double streetcar tracks, and congested—not only by motor and streetcar traffic but also by the multitude of horse-drawn carts of the "street A-rabs," or "hucksters," which were still

In earlier days, the long No. 23 suburban line followed Eastern Avenue to Essex and Middle River, but it was cut back to Back River in 1934 and abandoned completely in 1942. In its last days, a Back River car follows a weedy single track along two-lane Eastern Avenue. J. K. WINSLOW

determinedly part of Baltimore's culture. Many streets still had brick or stone-block paving. Synchronized traffic lights were almost nonexistent, and parking regulations were considerably more casual than they are now. (On the plus side, there were many fewer traffic lights and more parking spaces.) In all, local driving was a creeping, confused, and often bumpy ordeal.

To be sure, a few bold efforts already had been made to speed the ever-growing flow of motor traffic through Baltimore's archaic street layout. The Orleans Street viaduct had been open for nine years, and the Howard Street extension and bridge, which created a much-needed new north-south route, was five years old. Significantly, neither of these improvements included streetcar tracks. In fact, two car lines had been abandoned and several others rerouted so as not to hinder motor traffic on these new corridors.

Like most cities before the expressway era, Baltimore was basically more compact. People clustered closer to the city's center, and many more worked downtown. The downtown they traveled to, usually by streetcar, was a different downtown, too. Large or memorable buildings were few. There was no Charles Center, of course, nor any of the later tower buildings, and no Baltimore Arena, Convention Center, or Inner Harbor development. And, of course, there were no downtown stadiums; in fact, the now-departed Memorial Stadium did not even exist as such in 1945. In their place, the downtown was a haphazard jumble of dark brick buildings, most of which dated back forty years or more, the successors to buildings destroyed in and the survivors of Baltimore's most traumatic event, the Great Fire of 1904. In truth, to many outsiders and some natives, too, much of the city's downtown had an unimpressive, seedy look, particularly as one went south toward what is now called the Inner Harbor—then known as the "Basin"—and the Camden Station area. The average person noticed only a few memorable landmarks—the Bromo-Seltzer Tower, the Tower Building, and the dazzling art deco Baltimore Trust Building, as it was originally called (in 2002 the Bank of America building). Most of the rest seemed to merge into a clutter of undistinguished low-rise commercial structures that surrounded a past-its-prime Basin.

Or so it seemed at the time. Ironically, some are considered lost treasures now, notably the collection of nineteenth-century iron-front ware-

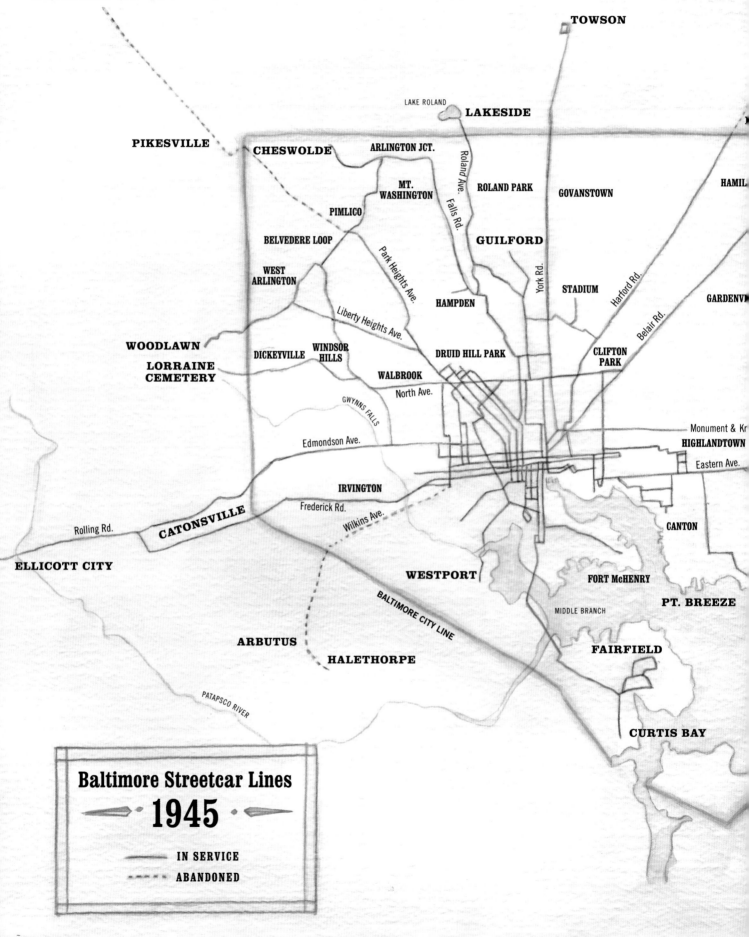

to REISTERSTOWN & GLYNDON

TOWSON

LAKE ROLAND

LAKESIDE

PIKESVILLE

CHESWOLDE

ARLINGTON JCT.

MT. WASHINGTON

ROLAND PARK

HAMIL

PIMLICO

GOVANSTOWN

BELVEDERE LOOP

Roland Ave.

Falls Rd.

GUILFORD

WEST ARLINGTON

Park Heights Ave.

STADIUM

York Rd.

GARDENV

Liberty Heights Ave.

HAMPDEN

Harford Rd.

Belair Rd.

WOODLAWN

DICKEYVILLE

WINDSOR HILLS

DRUID HILL PARK

CLIFTON PARK

LORRAINE CEMETERY

WALBROOK

North Ave.

GWYNNS FALLS

Monument & Kr

HIGHLANDTOWN

Edmondson Ave.

Eastern Ave.

IRVINGTON

CANTON

Rolling Rd.

CATONSVILLE

Frederick Rd.

Wilkins Ave.

ELLICOTT CITY

WESTPORT

FORT McHENRY

PT. BREEZE

BALTIMORE CITY LINE

MIDDLE BRANCH

ARBUTUS

FAIRFIELD

HALETHORPE

PATAPSCO RIVER

CURTIS BAY

Baltimore Streetcar Lines
· 1945 ·

——— IN SERVICE
- - - - ABANDONED

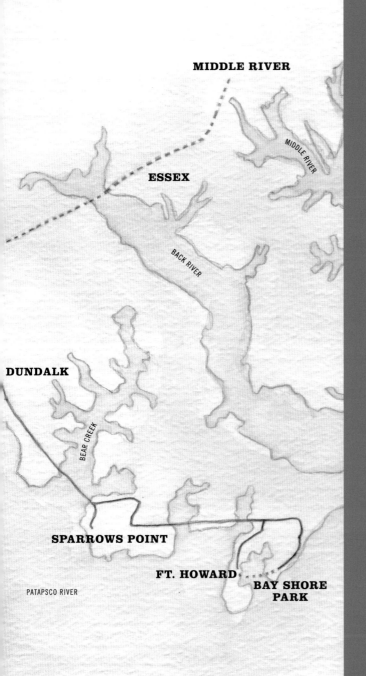

LE

VERLEA

MIDDLE RIVER

MIDDLE RIVER

ESSEX

BACK RIVER

DUNDALK

BEAR CREEK

SPARROWS POINT

FT. HOWARD

BAY SHORE
PARK

PATAPSCO RIVER

houses serving the teeming wholesale produce district east of Camden Station. But then, Baltimore merely felt a bit embarrassed by the large amounts of masonry flotsam left over from its more vigorous youth.

Regardless of what downtown looked like, it was well populated. It had little competition. With no expressways, there were no suburban office parks or shopping malls. True, a few suburban shopping centers had appeared, but they were close-in, cozy, and neighborly. Edmondson Village, the first "large" outlying center, was still two years away. For any serious shopping, people came downtown to the cluster of big department stores at Howard and Lexington Streets and the specialty shops along Charles. These were not just the center of the retailing world, they *were* the world. Suburban branches were unknown, and suburban competition was virtually nil.

People also usually came into town for first-run movies, and in this pretelevision time they certainly saw more movies of all kinds. At least eight major theaters sat close to the Howard-Lexington hub. Grandest was the Stanley on Howard Street, followed closely by Keith's and by Loew's "twin" Century/Valencia on Lexington Street and by the Hippodrome on Eutaw. (One other movie palace, the Town on Fayette Street, oddly was doing temporary duty as a parking garage.) Ford's Theater on Fayette fairly well monopolized the "legitimate" theater productions. And there were neighborhood movie theaters everywhere—sometimes two or three in the same block—including such sumptuous newcomers as the Senator on York Road and the Ambassador on Liberty Heights Avenue.

Out-of-town travelers usually stayed downtown, too. In 1945, the word *motel* still conjured up a picture of tacky little cabins along Route 1 to Washington or Route 40 to the northeast. Most Baltimore visitors headed for one of the four big city hotels—the Lord Baltimore, which was the businessman's favorite; the Emerson; the Southern; or that durable dowager, the Belvedere. The Mount Royal, near Penn Station, also had its following, although it was somewhat remote from the city's heart. The atmospheric old Rennert had been demolished four years before, but by then it had ceased to be a significant part of the commercial hotel scene.

Travelers, both natives and outsiders, had a lively and diverse transportation system from which to choose. Long-distance trips were usually made by train, of course. Baltimore's two major railroads, the Baltimore & Ohio and the Pennsylvania, competed across the board with services to New York, Chicago, Saint Louis, Detroit, and most other eastern and midwestern points. But for Baltimoreans, the B&O naturally was the preferred route unless some compelling reason intervened. When going to New York, the Pennsylvania's electrified "big red subway" was more frequent, faster, and more convenient, yet the B&O's genteel Royal Blue had an obstinately loyal, regular clientele.

Depending on where they were going, train travelers used no fewer than six downtown terminals. Most boarded at the B&O's Camden or Mount Royal Stations, or the Pennsy's Penn Station. But some commuters and people bound for certain local points also went to Calvert Station to catch the Pennsy's Parkton locals or to Hillen Station, where the Western Maryland Railway left for Glyndon, Westminster, and Hagerstown. (The Parkton locals and the Western Maryland trains also stopped at Penn Station, where many passengers boarded.) Finally, there was the Maryland & Pennsylvania Railroad's anonymous little shed below the viaducts at North Avenue and Howard Street. The "Ma & Pa," as everyone called it, always lived in a backwater of the railroad world but did offer a scenic, leisurely ride to places like Towson, Glenarm, Bel Air, Delta, and York. Almost nobody made the trip all the way to York, though, because the Pennsy got there much faster.

The Washington, Baltimore & Annapolis high-speed electric line had been gone for ten years, but its big terminal at Howard and Lombard Streets still stood, complete with platforms, tracks, and a large wall sign advertising WB&A services. The WB&A's surviving segment, the Baltimore & Annapolis Railroad, continued to run its heavy, dark red electric trains from Camden Station to Bladen Street in Annapolis. The B&A's big cars were now showing signs of age and neglect, but they still moved along fast: they took only fifty-five minutes from Baltimore to Annapolis, making all the numerous local stops. In the late twentieth century, their MTA bus replacements needed an hour and twenty-five minutes for the same run.

There were some other ways of traveling, too. Down at the Basin, the Old Bay Line's overnight Norfolk steamers departed from decaying but

Between 1935 and their demise in February 1950, Baltimore & Annapolis Railroad electric trains shared B&O's Camden Station, which bustled with steam- and diesel-hauled passenger trains for Philadelphia, New York, Washington, and major points west. On this afternoon in 1947, a B&A Annapolis train and a B&O Washington local leave simultaneously. The electric cars left the B&O main line near Monroe Street. A parking lot for offices at Oriole Park at Camden Yards now occupies this area. H. H. HARWOOD, JR.

Baltimore's downtown scenery was very different in 1946 as a No. 6 car for Fairfield rumbled south on Light Street past the McCormick building. Old Bay Line steamboats for Norfolk loaded to the right, and what was originally called the Baltimore Trust Tower (left rear) was the city's major landmark. It still stands today, having gone through a succession of names and now overwhelmed by a sea of new high-rise boxes. The "Basin" (behind the buildings to the right) is now the Inner Harbor, and the motley buildings bordering it have been replaced by the sanitized Harborplace, where armies of tourists now populate what was once a seedy backwater of the city. G. J. VOITH

ornate piers on Light Street opposite the McCormick spice building. Also leaving from Light Street was the Pennsylvania Railroad's Love Point ferry, the survivor of a one-time combination ferry-railroad route from Baltimore to Salisbury and Ocean City on the Eastern Shore. Ever practical, the Pennsy had recognized the inevitable and converted the ferry primarily to an auto carrier. "Smokey Joe," as the Love Point ferry was called, took a leisurely two hours and twenty minutes to cross the Chesapeake Bay. If nothing else, it was relaxing. But Smokey Joe and the Sandy Point–Matapeake ferries were the only ways to "drive" to the Eastern Shore unless one went north "around the Horn," through Havre de Grace and Elkton. The Bay Bridge was still in the future.

On the other side of the Baltimore harbor near Dundalk, DC-3 airliners flew from Harbor Field, the municipal airport—what later became the Dundalk Marine Terminal. But airplanes were a fairly minor method of going places. Air travel was expensive, and Baltimore was not a major point in the trunk airline networks.

However you traveled out of town, you usually took the streetcar to get anywhere within the city. Autos were fewer, driving was more difficult, and almost everything worthwhile was on a car line anyway. There were fewer routes than in today's MTA bus system, but those of 1945 carried more people and ran more often.

Indeed, streetcars seemed to be everywhere. A total of twenty-nine separate lines still operated in 1945, including branches and shuttle services. Car tracks crisscrossed downtown in a tight thicket and reached out to such suburban spots as Ellicott City, Catonsville, Woodlawn, Cheswolde, Lake Roland, Towson, Parkville, and Overlea. To the southeast and south, they ran to Sparrows Point and Fort Howard (and in the summer to Bay Shore Park), Curtis Bay, Fairfield, and Westport. Earlier, the cars had roamed even farther: to Reisterstown and Glyndon, Pikesville, Halethorpe, Carney, and Middle River. But during the Depression years, these light-traffic rural routes had been switched to feeder buses.

The fare, incidentally, was ten cents—exact change not required. Cheap, you may think, but at that time a clerk in the B&O Railroad's general offices earned about eight dollars a day and the "$10,000-a-year-man" was the symbol of salaried success.

The streetcar system not only was big and busy but also had almost every sort of scenery, construction type, and operational method. Once outside the city center, many routes ran over private open track, which was both traffic-free and pretty. Sometimes the track was laid in grassy boulevard median strips, sometimes it followed the side of the road, and sometimes it struck out on its own through woods and fields. Examples were almost too numerous to catalog: Outer Edmondson Avenue, Gwynn Oak Avenue, and Harford Road by Clifton Park all had long stretches of roadside right of way. Roland Avenue, Liberty Heights Avenue, and Dundalk Avenue had varying types of center-boulevard reserved track. The Mount Washington–Pimlico and Windsor Hills–Dickeyville areas had a little bit of everything. Upper Saint Paul Street in Guilford had a trolley layout almost unique in the country, carefully designed to fit the aesthetics of this rarified community. The two tracks of the No. 11 line through Guilford to Bedford Square were split, with a separate track on each side of the roadway set in a private reservation and bordered by a well-trimmed hedge.

In fact, many lines were worth riding strictly to savor their scenery or operating peculiarities. The high-speed, high-capacity Dundalk–Sparrows Point route looked much like a main line railroad, with its straight, open track, block signals, multiple-car operation, long Bear Creek trestle, and industrial surroundings. In a somewhat similar environment, Point Breeze cars picked their way through a puzzle of railroad tracks and manufacturing plants in the Canton area. At the other extreme was the bucolic line to Dickeyville and Lorraine Cemetery, whose single track meandered through woods, along roadsides and streams, and through the center of a tiny early-nineteenth-century mill village. All of that, incidentally, was inside the Baltimore city limits. Ellicott City trolleys ran cross-country west of Rolling Road, cutting through a high hill and rumbling over the Patapsco River on a long, steel truss bridge. At the northern boundary of the city, streetcar riders also could commune with unspoiled woods, streams, and ponds on their way to Lakeside loop near Lake Roland. In yet another kind of world was the lone shuttle car plodding over four blocks of Union Avenue in Woodberry, carrying workers from Hampden and the Falls Road car line down the steep hill to the old mills in the Jones Falls valley.

Some parts of Baltimore were downright rural, with "country trolleys" to match. Here, an ancient Brill semiconvertible meanders alongside Wetheredsville Road in Dickeyville on its way to Lorraine Cemetery. J. K. WINSLOW

Then there were several spectacular bridges and viaducts. Most memorable was the Guilford Avenue "elevated," running for over three-quarters of a mile above Guilford Avenue between Lexington and Chase Streets. Almost as awesome was the Huntington Avenue trestle, which hurdled the Stony Run valley between the north end of Huntington Avenue and the intersection of Thirty-third and Keswick in Hampden. And at the southeastern corner of the system, Sparrows Point cars crossed the wide expanse of Bear Creek estuary on a long, low timber trestle punctuated by a steel swing bridge at its center. East of Sparrows Point, other wood trestles carried an extension of the line over inlets on its way to Fort Howard and Bay Shore.

Populating this streetcar Garden of Eden was a fleet of 1,040 operating trolleys, most of which were as atmospheric as the lines on which they ran. They were a mishmash of ages, designs, and color schemes, with almost a forty-year spread between the newest and oldest cars. Most were old.

Newest in the fleet were 275 sleek PCC streamliners dating from the 1936–44 period. Unquestionably the finest in trolley technology, these were the product of a joint street-railway industry group called the Electric Railway President's Conference Committee—thus, the initials *PCC*. The PCC cars were smooth, fast, and quiet, at least by streetcar standards. All were originally painted in an attractive deep blue-green and cream combination with orange trim, designed by students at the Maryland Institute, College of Art.

Almost as young in years, having been built in 1930, but less advanced or as aesthetic as the PCCs, were 150 steel cars known in the trade as *Peter Witts*. The odd name honored a Cleveland public official who had designed their floor layout and original fare collection system. Their bright yellow and cream paint, set off with a red band, helped relieve their rather severe, squared-off look.

But the dominant Baltimore streetcar was a wood-bodied, ancient appearing ark called the *Brill semiconvertible*. (The name derived from a design that allowed the large side windows to be moved up into the roof area during the summers to provide fresh breezes.) Its classic turn-of-the-century design dated to 1905, although the ever conservative company management continued ordering them until 1918—

A wet snow pelts a PCC streamliner on one of Baltimore's last two car lines as it begins its run at Walbrook Junction on February 15, 1958. By the next morning, the snow was fifteen inches deep, virtually paralyzing this semisouthern city. F. W. SCHNEIDER III

by which time almost nine hundred were populating the streets. By 1945, they were clearly obsolete, both mechanically and visually, but they were also charming, provided that you did not have to ride them regularly. Indeed, to some outsiders they helped reinforce Baltimore's image of decaying gentility. The old Brills came in two color schemes; these told riders in advance which end of the car to board. Red and cream indicated a two-man car staffed by a motorman and conductor; passengers entered at the rear, where the conductor was stationed. Cars painted a pleasant yellow and cream, with thin, chaste red striping, were one-man cars; riders boarded at the front and paid their fares to the motorman. Some of the red two-man cars were fitted for train operation and often ran in two- and three-car sets on the heavy Sparrows Point line, where they were affectionately known as the "red rockets." The Brill semiconvertibles may have been old, but they were hardy and numerous. More than 570 of them still whined and clattered over the streets in 1945, and you were likely to see them on almost every line at one time or another.

Rounding out the roster were forty-three odd two-unit articulated cars, one of the very few fleets of articulateds ever regularly used on American street railways. At first glance, these looked much like the Brill semiconvertibles, and they were painted in the same yellow and cream colors, but they were unmistakably their own breed. Looking like a two-car train that wasn't fully hatched, they were impressive and fascinating to watch as they swallowed up crowds or negotiated curves. The articulateds had been home-built from old semiconvertibles and obsolete open-car bodies, an inventive way of getting extra rush-hour capacity cheaply. But by 1945, their ranks were thinning and the operable survivors were confined to only four or five lines, and then only at various peak hours. (Interestingly, although very few traditional American streetcar companies ever adopted the articulated design, it is now commonly used on modern light rail systems—including Baltimore's.)

Motor buses were in evidence, of course, but they were in the minority, and you saw surprisingly few of them in the downtown area. Compared to the fleet of more than one thousand streetcars, there were only 242 motor buses, and in 1945 there were only three trunk bus routes. Most of the remaining lines were "feeders," radiating from streetcar termi-

Contemporaneous with the Model A Ford, Baltimore's so-called Peter Witt cars shared the same utilitarian design philosophy. By 1930 standards they were fast and comfortable—and certainly an exhilarating change from the omnipresent, archaic semiconvertibles. These, in fact, were the first "modern" steel cars ordered in any quantity by the Baltimore system, which finally did so at least fifteen years after most other streetcar companies had adopted steel construction. Originally, passengers entered at the front, left by the center door, and paid when they passed a conductor stationed by the rear door. Typical of Baltimore's Peter Witts is the 6150, posed at the Belvedere carhouse in 1946. L. W. RICE

nals or transfer points to reach newly developed or lightly populated outlying areas. These buses served such spots as Pikesville, Reisterstown and Glyndon, Randallstown, Homeland, Northwood, Carney, and the war-created housing developments at Middle River and Armistead Gardens. Riding the feeder buses usually was a bit of a nuisance, since you generally had to transfer to or from a streetcar to get anywhere. For example, anyone going into the city from Homeland rode a Route "O" bus to Bedford Square and changed to the No. 11 streetcar. Northwood residents took the "T" bus to Thirty-third and Greenmount, where they waited for a No. 8 car, which they hoped was not too crowded. Several independent bus companies also operated to various suburban points from downtown or from the ends of car lines.

A third variety of transit vehicle in Baltimore was the electric trackless trolley. These bus-trolley hybrids were very smooth, quiet, and, all things considered, probably the most pleasant to ride of anything in Baltimore, past or present—at least when on smooth pavement. In 1945, 128 trackless trolleys operated over three routes, all of them one-time streetcar lines. More would come soon to replace other car lines, but, sadly, all would be gone even before the last streetcar ran.

The events leading up to that last streetcar started in 1945 and moved quickly thereafter. Although Baltimore's transit scene in 1945 was active and atmospheric, there were many problems below the surface. The worst was money. Although Baltimore had been a premier streetcar city, the system was never very profitable. It was then operated by the Baltimore Transit Company (BTC), a private concern that in 1945 was only ten years old. BTC was the Depression-born reincarnation of the bankrupt United Railways & Electric Company, originally formed in 1899 to consolidate and modernize Baltimore's diverse independent streetcar companies. Prosperity had chronically eluded the "United," and thus far BTC had been no luckier. It was now struggling with an aging streetcar fleet and obsolete track layout, both of which had been badly battered by the wartime crush.

At the same time, the transit company faced the need to serve the growing areas beyond and between the car lines and generally to adapt its route system to a changing city. With the war's end, patronage quickly resumed the long downward slide that had started twenty years before.

The PCC car was probably the finest streetcar design ever built and was the most modern equipment in Baltimore—and many other cities. Baltimore bought 235 of them from 1936 to 1944. No. 7344, posed here in September 1941, had just been recently delivered and is resplendent in its original regal paint scheme of dark green and cream with a thin orange stripe below the window line.
H. H. HARWOOD COLLECTION

The drastic drop in ridership created a cash flow crisis just as modernization was most necessary. It seemed clear that the economics of streetcar operation were questionable, even if the capital could be found to rebuild the lines and replace the cars.

Two events speeded the end of the streetcars. First, in 1945, working control of Baltimore Transit was taken over by Chicago-based National City Lines, a bus-minded holding company that specialized in modernizing ailing trolley operations. Second was the arrival of Henry Barnes, the creative, aggressive, and controversial traffic commissioner hired to unclog the city's snarled streets. Barnes's concepts of "free-flow" (for autos), with one-way street systems and expanded traffic lanes, simply were incompatible with tracks and trolleys.

Buses seemed the solution to everyone's problems. Clearly, they were more flexible, were considerably cheaper to buy and run, and required no track and wire maintenance expenses or commitments to help repair or replace paving. Equally important was that they offered an excellent hedge against the uncertainties of declining business and shifting riding patterns. It was, after all, a generation before OPEC or air pollution concerns.

And so, on June 22, 1947, the first wave of postwar bus substitutions began. On that date four north Baltimore lines ended, victims of the conversion of Saint Paul, Charles, and Calvert Streets to one-way traffic. More changeovers came quickly. Eleven years later only two trolley routes remained: the heavily used No. 8 Towson–Catonsville and No. 15 Overlea–Walbrook Junction lines. Their passenger volume and the lack of money for bus replacements allowed a brief reprieve, but the inevitable finally came in the bleak predawn hours of Sunday, November 3, 1963. In the meantime, all those expressways, malls, office parks, and urban redevelopments were under way or already completed.

Not only was Baltimore no longer a streetcar city, but it never could be again.

Ancient but impressive, the "red rocket" trains were rolling Baltimore landmarks. Headed by 5181, this pair rumbles past the city's center, Baltimore and Charles Streets, during a Labor Day excursion for streetcar fans in 1948. Directly to the left is B&O's general office building; to the rear is the locally revered O'Neill's department store, now replaced by the One Charles Center building.
G. J. VOITH

Downtown Baltimore was a confusing and noisy mishmash of rails, crossings, and switches. Virtually every street carried car tracks, and a wide variety of routings and loops was possible. Each line seemed to follow a different path through the area, with some terminating downtown, others running through and turning back on the outer fringes, and still others simply running through. The most famous example of the latter was the No. 8 Towson–Catonsville line, which began on York Road in Towson, passed through the city center, and then followed Frederick Road to Catonsville.

The No. 18 Canton–Pennsylvania Avenue line ran through the downtown area, using Lombard and Saratoga Streets. Here semiconvertible No. 5742 swings west onto Saratoga Street from Eutaw in June 1951 on its way to Pennsylvania and North Avenues. A line maintenance crew rests at the curbside waiting for a lull in the streetcar traffic. E. S. MILLER

Heading outbound on Lexington Street at Park Avenue, a No. 14 Edmondson Avenue car for North Bend takes on passengers in June 1946. Keith's Theater was later razed for a parking garage, and its entranceway was remodeled into a store. The nondescript buildings behind the streetcar on Liberty have disappeared for Charles Center. Lexington Street itself is now a pedestrian mall. R.W. JANSSEN

ABOVE To the south, a car on the No. 33 line from West Arlington makes its terminal loop in front of the B&O's grimy Camden Station in 1946 just as a trackless trolley from Morrell Park bounces by on the stone paving. Although the hotel and other buildings on the left are gone, Camden Station not only was preserved but has been restored to its original post–Civil War appearance as part of the new baseball stadium project. It no longer serves railroad passengers, however. G. J. VOITH **BELOW** Four blocks east on Saratoga Street, PCC No. 7084 handles a No. 14 Edmondson Avenue run. It waits to turn south on Charles to complete a terminal loop that will take it back west on Lexington. The classic 1908 Metropolitan Savings Bank building on the left was sacrificed to the Charles Center redevelopment in 1963, but happily the other buildings in this 1949 scene have been preserved and restored. F. S. MILLER

Typical of vanished downtown Baltimore are the scenes on these two pages.

ABOVE | The First National Bank building, standing tall in the rear, is the only present-day landmark to place this 1950 scene of an outbound Garrison Boulevard PCC car running west on Redwood Street at Hopkins Place. All of the immediate surroundings have been removed and replaced with the Hopkins Plaza complex, while the photo itself is taken from the site of the Baltimore Arena. E. S. MILLER **TOP RIGHT** | Three blocks south on Hopkins Place, where it becomes Sharp Street, this 1950 scene shows Peter Witt car No. 6125 prepared to turn west onto Camden Street. Now all of these surroundings have been obliterated by the Convention Center. E. S. MILLER **MIDDLE RIGHT** | Two blocks farther south, PCC car No. 7065 makes its way south on Sharp Street at Conway, heading for a loop at Light and Lee Streets. To the left in this 1950 view are the B&O's public freight delivery tracks

where much of the city's fresh fruits and vegetables arrived. The entire area behind the streetcar has now metamorphosed into the Convention Center. Still punctuating the Baltimore skyline is the perennial Bromo-Seltzer Tower. E. S. MILLER

BELOW RIGHT | Curtis Bay–bound Peter Witt No. 6047 is briefly stymied at Baltimore and Light Streets by someone's bright new 1947 Nash, which, apparently, cannot retreat. The car line was in its last year, to be replaced the following March by yellow-green and light gray General Motors (GM) buses like the interloper on the right. Thirty-five years later rails returned to this spot, but underground. It is now the site of the MTA Metro's Charles Center terminal. All the buildings behind the streetcar are now gone;

ABOVE | During the late 1940s, No. 26 Sparrows Point cars came into downtown on Lombard Street and looped at Pearl, west of the city's center. Heading east on Lombard, a pair of Sparrows Point "red rockets" clatters over the five-point intersection with Howard and Liberty Streets in May 1948. To the left was the Washington, Baltimore & Annapolis's Baltimore terminal. Although the last WB&A interurban had run thirteen years before, the name was kept alive by a restaurant in the building. R. W. JANSSEN

ABOVE RIGHT | Inbound semiconvertible No. 5853 rolls past a portable traffic control booth on Lombard at South Street in September 1949. At the time, several major downtown intersections were still manually controlled. This one rated only part-time police coverage, and the booth was rolled away during periods of light traffic. E. S. MILLER **MIDDLE RIGHT** | A two-car Sparrows Point train taxes the trusses of the 1877 Lombard Street bridge over the Jones Falls in 1950. The bridge long outlasted the cars, final

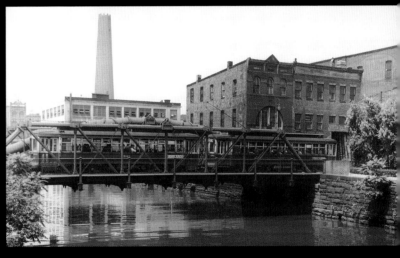

ly being replaced about 1973. In addition to vehicles, the unique bridge also carried a water main, visible behind and above the rear car to the left. Baltimore's most famous industrial relic, the 1828 Phoenix Shot Tower, pokes up in the rear, three blocks north. Molten lead was prepared at the top of the tower and dropped into water filled receivers at the bottom to form perfect spheres for shot. E. S. Miller   BELOW RIGHT │   Streetcars were nearing the end in this October 1963 view, but the durable Shot Tower still dominates the downtown scene. PCC cars running on lines No. 8 (near) and No. 15 (distant) simultaneously swing onto Fayette Street in the City Hall area. By then the complex trackwork at Holliday Street in the foreground was only partly used to turn back certain

ABOVE | An inbound No. 8 line PCC rolls down Hillen Street past the Western Maryland Railway's 1875 Hillen Station complex. When this 1950 photo was taken, the Western Maryland still ran two cleanly maintained Hagerstown locals and a commuter run to Union Bridge. Most passengers, however, boarded them at better-located Penn Station. The Victorian terminal complex came down in 1953, replaced by a mundane warehouse, and the WM ran its last Baltimore passenger train in 1957. E. S. MILLER

ABOVE RIGHT | Watched over by an ornate "bishop's crook" street lamp at its right, outbound articulated No. 8124 negotiates the switch at Hillen and Forrest Streets, carrying an afternoon rush hour crowd on the heavy Greenmount Avenue–York Road

MIDDLE RIGHT | Inbound on Gay Street, a dark green PCC on the No. 15 Belair Road line nears Lexington Street. Most buildings in this May 1950 scene were leveled for the Jones Falls Expressway extension, which now bridges Gay Street a block behind the car. One of the buildings on the left, however, still survives as a Baltimore railfan haven—the M. B. Klein hobby store. E. S. MILLER

BELOW RIGHT | The Baltimore Transit Company acquired a new, albeit short, piece of private right of way in 1941 when the Latrobe Homes urban renewal project swallowed up a block of Ashland Avenue between Aisquith and Ensor Streets. The street was closed, but the car line was continued on the old alignment within the new housing development. PCC No. 7355 heads into town at

Guilford Avenue Elevated

The famous Guilford Avenue "elevated" was the fastest, most direct, but least utilized of the major downtown routes. Sometimes called the "viadock" in Balmorese, it was originally built in 1892–93 by the Lake Roland Elevated Railway, the transit adjunct of the Roland Park residential development. (It had the distinction of being the first electrified elevated line in the United States.) As a latecomer to Baltimore's transit scene, the Lake Roland line was forced to use Guilford Avenue, which already was occupied by the Northern Central Railway (later the Pennsylvania Railroad) tracks reaching Calvert Station and various warehouses. Thus, the viaduct was necessary to hurdle the railroad activity. But the route lay on the fringes of commercial and residential areas, and for most its life it carried lighter-density car lines. In June 1947, the heavy No. 8 Towson–Catonsville line was rerouted onto the "el," but in January 1950 service reverted to Greenmount Avenue and the structure was closed and dismantled.

The south end of the "el" came to earth in the block between Saratoga and Lexington Streets. A PCC car descends the ramp on its long trip from Towson to Catonsville in September 1949. Surrounded by warehouses, the renowned House of Welsh lurks in the right rear. E. S. MILLER

The Guilford Avenue elevated's four-thousand-foot length is evident in the two views above and above right, both taken from the Orleans Street viaduct.

ABOVE LEFT | Looking south, a Towson-bound PCC car pulls away from the Pleasant Street station. Railroad sidings served many of the now-removed commercial buildings alongside. E. S. MILLER   **ABOVE RIGHT** | Looking northeast toward the city jail (at the right rear), an inbound No. 8 line car passes the rabbit warren of warehouses, railroad sidings, light industries, and a grain elevator that filled the area before the Jones Falls Expressway wiped them out. R. W. JANSSEN   **MIDDLE RIGHT** | From below the "el," it is clear why Guilford Avenue was hard to navigate on the surface. G. J. VOITH   **BELOW RIGHT** | The less-recorded north end of the viaduct was at Chase Street, seen in 1949 as PCC No. 7033 came cautiously down. On its way to Towson, it will use Guilford as far as North Avenue and then will turn east to Greenmount. E. S. MILLER

"El" stations were spartan affairs. Madison Street was typical. The Madison Street "el" station, shown here topside as a southbound car for Irvington glides into the station. The fan-shaped iron barrier in the foreground was meant to discourage anyone from short cutting over the girders, which was tried at least once, with a fatal result. E. S. MILLER

A few lines skirted the downtown area but did not enter. The heaviest route was the No. 13 North Avenue crosstown line running from Walbrook to Gay Street and crossing a multitude of radial routes in the process. Several other car lines used sections of North Avenue on their way in and out of downtown.

LEFT | Eastbound on the No. 13 North Avenue crosstown line, PCC car No. 7414 comes off the North Avenue viaduct at Howard Street. The stairway behind the 1947 Packard on the left led down to the "Ma & Pa" railroad terminal, a sma ter little shed that he l

sufficed since the railroad's attractive stone street-level station had been demolished in 1937 to make way for the Howard Street extension and bridge. In the haze behind the auto is the also-departed Mount Royal water pumping station, sacrificed for the Jones Falls Expressway. E. S. MILLER   **ABOVE RIGHT**   Typical of North Avenue transfer points was Linden Avenue. The viewer looks west in September 1949 as an inbound PCC from Woodlawn loads. Urban renewal now has left this area mostly barren, although the one-story block on the right remains. The Linden Theater, however, closed and was converted to other uses. E. S. MILLER **BELOW RIGHT**   The North Avenue Market was built in 1928 for the uptown trade. No. 5682 working eastbound on the No. 13 line

ABOVE | Another "fringe" route was the No. 30 Fremont Avenue line, circling the west side of the city from North Avenue and Charles Street to South Baltimore. In this November 1949 view, PCC No. 7135 pauses at Fremont and Edmondson Avenues on its roundabout way to South Charles and Barney Streets. Painted on the street is a warning to motorists that the streetcars overhang when turning. E. S. MILLER **ABOVE RIGHT** | Brill semiconvertible No. 5584 is beginning its circuitous tour covering the range of Baltimore neighborhoods from Druid Hill Park to Guilford via downtown. The No. 11 line car has left Park Terminal and is heading south on Fulton Avenue and crossing Walbrook Avenue, just north of North Avenue. J. K. WINSLOW

BELOW RIGHT | Another of the several routes terminating at Druid Hill Park was the No. 16 Madison Avenue line, which eventually found its way to the foot of Broadway in Fells Point. With the park's impressive Madison Avenue entrance in the background, No. 5698 awaits a return run on Cloverdale Street in April 1948. The car line's life ended a week after the photo was taken. R. W. JANSSEN

Western Suburban Lines

Two lines reached west to Catonsville—the west end of the famous No. 8 Towson–Catonsville line, running via Frederick Road, and the No. 9-14 line along Edmondson Avenue, which continued to Ellicott City. Frederick Road was the busier, passing through the commercial centers of Irvington and Catonsville, but Edmondson Avenue was the prettier, with its tree-bordered private track serving well-tended

Irvington • Edmondson Village • Catonsville • Ellicott City

Row houses, trees, and streetcars—these were Baltimore in the first half of the twentieth century. In October 1963 the trolleys were in their last month, as battered PCC No. 7418 makes its way east on Edmondson Avenue at Schroeder Street. The car is running on a gerrymandered No. 15 Overlea–Belair Road–Walbrook Junction line, patched together in 1954 from parts of the one-time No. 4 Edmondson Avenue–Walbrook Junction and No. 15 Belair Road–West Baltimore Street lines. F. W. SCHNEIDER III

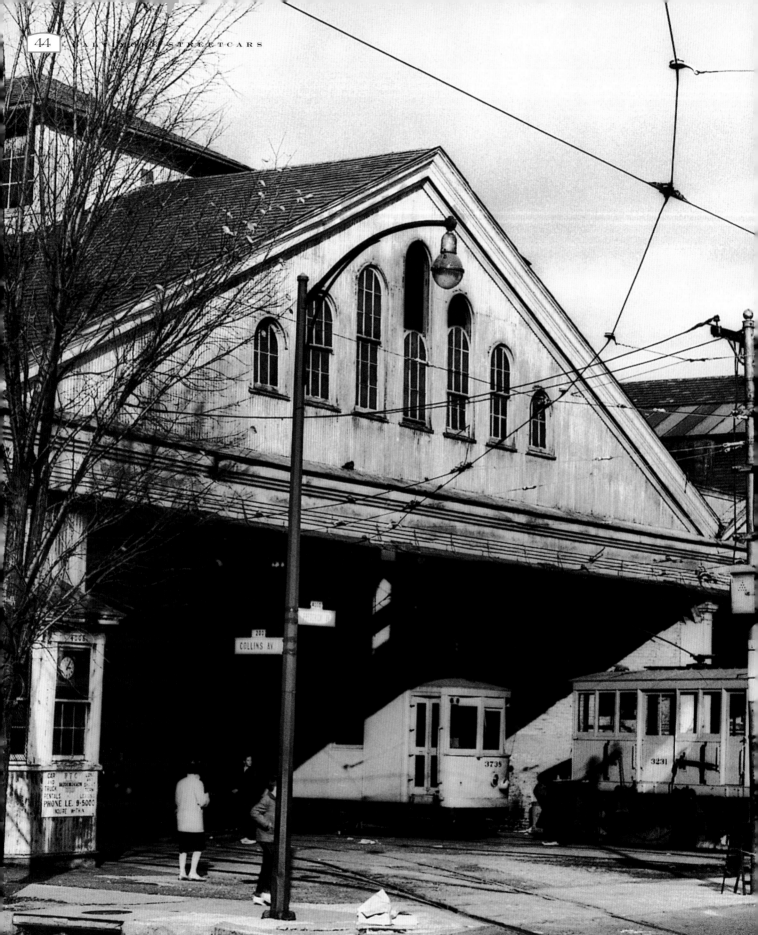

A major operating center for the Frederick Road line was this 1898 carhouse at Collins Avenue in Irvington, shown at the end of its life in November 1963. (Note the snow sweeper poking its nose out from the center of the barn, ready for the snows that, for it, will never come.) The old carhouse outlived many of its more modern brothers on the Baltimore system and was one of the last two in service when the last streetcar ran. Sadly, it was subsequently demolished. H. H. HARWOOD, JR.

ABOVE | On its way to Catonsville amid typical Frederick Road surroundings in May 1951, No. 7380 passes Shady Nook Avenue, just east of the present Beltway interchange. E. S. MILLER   **ABOVE RIGHT** | West of the center of Catonsville, the original No. 8 line alignment followed the north side of Frederick Road on a private right of way. As part of a 1951 street-widening project, the tracks were relocated to the center of the road, in what was to be one of the last major streetcar line reconstruction projects. In this scene at Stanley Drive in May 1951, the job was half done. An eastbound No. 8 PCC car rolls over the new line, while the old westbound track is barely visible on its alignment at the far right. E. S. MILLER

MIDDLE RIGHT | For its finale, the No. 8 line swung north from Frederick Road near Montrose Avenue and plunged into the woods for a brief run to its terminal at Edmondson Avenue. Towson-bound No. 7364 is heading south, nearing Frederick Road in August 1963. H. H. HARWOOD, JR. **BELOW RIGHT** | The Edmondson Avenue route generally passed through more newly developed areas, and in the late 1940s Baltimore's newest and most exciting development was the "big" Edmondson Village shopping center, seen here in January 1951. Eastbound No. 7081 pauses opposite the center's Georgian Revival–style theater. The line along this sec-

ABOVE | A short distance to the west, the countryside along Edmondson Avenue was almost unspoiled. Here, on the same early spring day in 1951, a semiconvertible bound for Ellicott City passes Harlem Lane. The area directly ahead of the car is now the interchange with the Beltway. E. S. MILLER   **ABOVE RIGHT** | West of its junction with Baltimore National Pike, Edmondson Avenue turned into a residential street and the car line followed suit—operating first on a tree-lined private right of way set between two lanes of a boulevard-style road and then on the south side of a pleasant two lane roadway. This outbound semiconvertible is emerging from the tree-lined "boulevard" section of Edmondson Avenue east of the North Bend

MIDDLE RIGHT | It is early in April 1951, and greenery is sprouting along with the Academy Heights development. Heights home buyers could catch the No. 14 streetcar for town, such as PCC No. 7377 swinging around the bend of Edmondson Avenue at Nunnery Lane. E. S. MILLER

BELOW RIGHT | Edmondson and Ingleside Avenues, on the outskirts of Catonsville, looked like a country crossroads in 1951. Car No. 7378 is outbound for the Rolling Road loop. Just around the curve to its rear is the present-day Beltway interchange. E. S. MILLER

ABOVE | An Ellicott City–bound car pauses briefly at the Rolling Road loop waiting station before plunging ahead into the woods. Cars of the No. 14 line turned back at this loop terminal, located just west of the Edmondson Avenue and Rolling Road intersection and built in 1941. This section of the old trolley right of way has since been paved over as an extension of Edmondson Avenue. E. S. MILLER **ABOVE RIGHT** | Edmondson Avenue skirts the north side of "old" Catonsville, then as now an area of large homes and lots of trees. In this view near Glenmore Avenue, PCC car No. 7377 is returning from Rolling Road to its downtown terminal at Charles and Lexington Streets. In the far distance, an Ellicott City car follows it into town. E. S. MILLER **MIDDLE RIGHT** | Edmondson and Dutton Avenues, locally called Catonsville Junction, was a small commercial center serving northwest Catonsville and a key street-

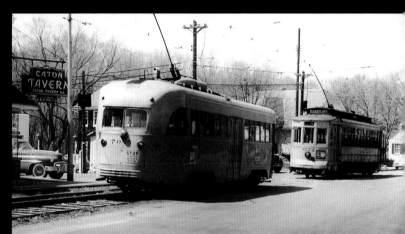

car transfer point. Out of the picture on the left was the end of the No. 8 Towson–Catonsville line. Passengers between Frederick Road points and Rolling Road, Oella, and Ellicott City changed cars here. Some Ellicott City runs also switched back here; downtown riders took either No. 8 or No. 14 cars. The bucktoothed 1950 Buick by the Caton Tavern was one year old as a Rolling Road PCC car followed an ancient Ellicott City semiconvertible west. After the demise of the Edmondson Avenue line in September 1954, isolated Ellicott City shuttle cars continued running from this point for another ten months before they, too, died. E. S. MILLER

BELOW RIGHT | Between Catonsville Junction and Rolling Road, the car tracks occupied the center of Edmondson Avenue. Nearing its terminal at Rolling Road, this PCC car is skimming westward from the junction, which can be vaguely seen in the right rear. J. K. WINSLOW

The farther west the Ellicott City car went, the deeper the woods got—or so it seemed. This was typical of the scenery as the line dropped into the Patapsco River valley. The car is westbound in April 1951. E. S. MILLER

Ellicott City, a late-eighteenth-century flour-milling town, is nestled in a gorge of the Patapsco River fourteen rail miles west of Baltimore—but it really exists in another world. The B&O Railroad's pioneering strap rails reached the town in 1831 by following the river valley. The cross-country trolley line arrived in 1899.

ABOVE | Cars crossed the Patapsco River on this long truss bridge, which actually carried a gauntleted double track (two tracks overlapping one another). The Frederick turnpike crossed immediately out of the picture on the right. On the far riverbank are ancient mill workers' houses and the car line's wilderness right of way uphill through the woods to Catonsville. Ellicott City's

Main Street lies ahead of semiconvertible No. 5767 as it rolls off the bridge into town in April 1951. E. S. MILLER

ABOVE RIGHT | Crammed into a narrow tributary valley at a right angle to the Patapsco River, Ellicott City has always had a peculiarly European look, with its steep hillsides, narrow streets, and closely spaced stone buildings. In a 1947 view west from the B&O Railroad bridge, the red and cream No. 5292 grinds up Main Street to its Fells Lane terminal, pursued by a rumble-seated roadster. R. M. VOGEL **BELOW RIGHT** | Ellicott City cars followed narrow Main Street all the way through town and uphill to its west end. At the end of the line, there was no room to lay over on the street itself, so cars turned off into a small off-street terminal at Fells Lane. Semiconvertible No. 5706 approaches the curve leading into the terminal in April 1951. E. S. MILLER

Walbrook, at the west end of North Avenue, started as a late-nineteenth-century suburban development but quickly grew after it became

Walbrook • Windsor Hills • Dickeyville • Lorraine Cemetery • Ashburton
Gwynn Oak Park • Woodlawn • West Arlington • Belvedere • Park Heights

It is a quiet Sunday noon in the center of Walbrook as No. 7409 heads east on North Avenue in January 1951. At this time three neighborhood theaters populated this block: the Walbrook on the right and the Windsor and the Hilton on the left. The Windsor and Hilton both closed in the 1950s, while the grandiose Walbrook, built in 1916, lasted until 1964. All three still stand, converted to other uses, the Walbrook as a church. E. S. MILLER

Walbrook Junction, four blocks northwest at Clifton Avenue and Garrison Boulevard, was an active terminal and transfer point. In the 1940s, four lines came together here: No. 4 Edmondson Avenue (which continued to Windsor Hills), No. 13 North Avenue crosstown, No. 31 Garrison Boulevard, and the picturesque No. 35 line to Lorraine Cemetery via Dickeyville. This view looks northwest at the Clifton and Garrison intersection in mid-1950. To the right, green No. 7001, Baltimore's first PCC car, is inbound from Garrison Boulevard. To the left, No. 7419, wearing a National City Lines "fruit salad" yellow-green and light gray paint scheme, has made its loop at the junction terminal and prepares to start east on a North Avenue run. E. S. MILLER

The Walbrook Junction loop itself lay in the triangle bounded by Clifton Avenue and Windsor Mill Road. The viewer looks along Windsor Mill as another North Avenue car makes ready to leave. E. S. Miller

ABOVE | No. 5622 is inbound for Walbrook Junction on Clifton at Mount Holly. E. S. MILLER   **ABOVE RIGHT** | West of Walbrook Junction, the No. 35 Lorraine line shared track with the No. 4 line as far as Windsor Hills, following Clifton Avenue through tree-shaded residential areas. In an archetypical streetcar scene, semiconvertible No. 5680 is outbound on Clifton Avenue at Allendale Road in July 1950. E. S. MILLER   **BELOW RIGHT** | Windsor Hills was a turn-of-the-century "garden suburb," laid out with winding streets, large houses, and lots of woods. The trolley fitted right in. No. 5622 is westbound for Lorraine at Clifton and Queen Anne Road in July 1950. The line itself was opened in 1904, contemporaneous with the development. E. S. MILLER

Beyond Windsor Hills, the Lorraine line got progressively more rural—and went farther back in time. The following series of views, all taken between 1948 and 1950, trace the route between Windsor Hills and Dickeyville. Although decidedly a minor operation in the vast Baltimore system, it was the most photogenic and generally charming in this city or almost any other; as a result it got photographic attention far beyond its importance.

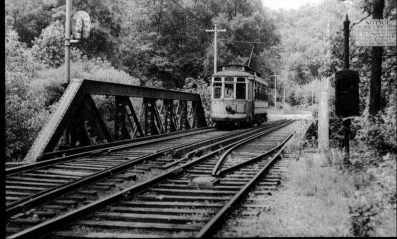

MIDDLE RIGHT | Near Fairfax Road, deep woods surrounded the cars. A light snowfall dampens the clatter of No. 5395 as it returns uphill from the Gwynns Falls and Dickeyville. R. W. JANSSEN

BELOW RIGHT | Double track resumed briefly at the little truss bridge over the Gwynns Falls, the stream that once provided power for mills that first appeared in this area in 1762. This view looks west toward Wetheredsville Road (Dickeyville's earlier name) as eastbound No. 5622 waits to meet a westbound car at the end of the double track. E. S. MILLER

ABOVE | Trolley riders were taken back more than one hundred years as the cars entered Dickeyville, a tiny mill town that grew up in the late 1700s and early 1800s and then froze in time. It still remains an isolated gem, superbly restored, inside Baltimore's city limits. In this atmospheric scene, which could be New England circa 1905, eastbound No. 5727 follows Wetheredsville Road as a mill worker returns home in midafternoon. J.W. VIGRASS **ABOVE RIGHT** | West of the bridge over the Gwynns Falls, the trolley track played tag with Wetheredsville Road, swinging from the south side to the center to the north side as it approached Dickeyville. No. 5622 is on its way to Walbrook Junction in July 1950. E.S. MILLER

MIDDLE RIGHT | "Downtown" Dickeyville in May 1949 as a railfan charter rests in the stub-end siding next to the community's main waiting station. In earlier years, cars from downtown terminated here and a jerkwater shuttle ran to Lorraine Cemetery. R. W. JANSSEN **BELOW RIGHT** | Beyond Dickeyville, the Lorraine route described a series of broad S curves as it worked its way alongside Forest Park Avenue, Windsor Mill Road, Kernan Drive, and Dogwood Road to reach the cemetery. Here, inbound No. 5622 swings into the curve connecting Windsor Mill and Forest Park, passing the gates of Kernan Hospital, a famous children's ortho

| The outstanding landmark on the Woodlawn line—and a major Baltimore tradition for more than seventy years—was Gwynn Oak Park. Set in a sylvan spot along the Gwynns Falls, the park opened in the 1890s and was owned by the streetcar company for much of its life. In the Woodlawn line's last summer, yellow-orange Peter Witt car No. 6119 passes the park's roller coaster in August 1955. Gwynn Oak Park finally closed in 1974 and burned the following year, yet No. 6119 still survives at the Baltimore Streetcar Museum. G. J. VOITH   ABOVE RIGHT | Locally known as Gwynn Oak Junction, the intersection of Liberty Heights and Gwynn Oak Avenues was the dividing point for branches to Woodlawn (right) and West Arlington–Belvedere (left). Bracketed by the

1933 Gwynn Theater and the moderne-style 1935 Ambassador Theater, outbound Peter Witt No. 6120 takes the switch for Woodlawn. The two theaters are long since closed, the Gwynn shortly after this April 1951 photo was taken. Both remain in other uses. E. S. MILLER **BELOW RIGHT** | Baltimore's last major streetcar trunk line extension was the route out Liberty Heights Avenue from Reisterstown Road to Gwynn Oak Avenue, opened in 1917. For most of its life, it carried No. 32 Woodlawn cars, although after June 1948, the West Arlington line also was routed into town over Liberty Heights Avenue. On its way to Woodlawn, a PCC car passes Denison Street, west of Lake Ashburton, in April 1951. E. S. MILLER

LIBERTY 2911   LIBERTY 2912

C-AND-K
SUPERETTE

FRESH
FRUITS
ETABLES
ROSTED
OODS

FRESH
MEATS
AND
POULTRY

Crush

PRESSING
WHILE
YOU WAIT

FROZEN FOODS   MEATS   GROCERIES

TABLES   FRUITS   MEATS   GROCERIES

324
ALL 515

7068

PHIL
BAKER

ABOVE LEFT | **West Arlington was another turn-of-the-century residential development, mostly single homes, hedges, and trees.**

This view is at Gwynn Oak Avenue at Post Road in April 1951. The PCC will go downtown via Liberty Heights Avenue. F. S. MILLER

Returning to Gwynn Oak Junction, we now look along the West Arlington branch. Originally designated No. 5-33, this line came out of town on Park Heights Avenue and terminated here at Liberty Heights and Gwynn Oak Avenues. In June 1948, the Park Heights line died, and West Arlington cars were routed into town on Liberty Heights. The junction was extra busy in April 1947 as fan excursion car No. 5566 shared tracks with two other semiconvertibles at the terminal switchback. L. W. RICE

The park and the broad Gwynns Falls valley can be seen to the left of Woodlawn-bound PCC No. 7067 running alongside Gwynn Oak Avenue near Poplar Drive in April 1951. From here to Woodlawn the line originally had been single track

ABOVE | On their way between Gwynn Oak Junction and Belvedere Avenue, West Arlington cars used some stretches of private right of way to cut through areas where there was no direct street routing. This was one such spot. G. J. VOITH

ABOVE RIGHT | Between Reisterstown Road and the present Wabash Avenue, the West Arlington and Garrison Boulevard cars crossed the Western Maryland Railway main line on another "cut-through" private right of way that would later become Belvedere Avenue. In this 1947 scene, No. 5757 has received a green signal at the protected crossing and is about to crash across on its way to West Arlington. The cars in the distance are waiting to cross Reisterstown Road. This section is now paved, and directly behind the

| To the south, in the Forest Park area, an inbound Garrison Boulevard PCC car has just crossed Liberty Heights Avenue and is approaching the Berwyn Avenue intersection in May 1950. Before the inner section of the Liberty Heights trackage was opened in 1917, Woodlawn cars came out Garrison Boulevard to this point and continued on Berwyn (seen to the left) to Liberty Heights. E. S. MILLER

HOT LUNCHES

Coca-Cola

SANDWICHES

The Belvedere carhouse, on Belvedere Avenue east of Reisterstown Road, was a thriving hub for northwest Baltimore car and bus lines. As late as 1948, three streetcar routes and two feeder bus lines terminated here. Across the street once was long-defunct Electric Park (an amusement park) and later a housing development.

Belvedere Avenue east of Reisterstown Road looked like this in May 1950, as a PCC started west on a Garrison Boulevard run while Peter Witt No. 6021, assigned to the Mount Washington–Pimlico line, relaxes by the ornate 1907 carhouse. E. S. MILLER

MIDDLE RIGHT | Another Northwest Baltimore streetcar corridor was Reisterstown Road and Liberty Heights Avenue, which carried No. 5 and No. 33 cars through Park Circle and on to Pimlico. Before 1932, the No. 5 line extended all the way to Reisterstown and Emory Grove via Park Heights, Slade Avenue, and Reisterstown Road, but later the line was progressively cut back to Park Heights and Manhattan Avenue, just beyond present-day Northern Parkway. Here an ancient Brill semiconvertible grinds out Reisterstown Road south of Park Circle in June 1948. R. W. JANSSEN   **BELOW RIGHT** | No. 5747 from West Arlington picks through the pedestrians as it turns from Belvedere Avenue onto Park Heights to head downtown. The Mount Washington–Falls Road Peter Witt behind will go straight across the intersection and past Pimlico racetrack. G. J. YOITH

Easily Baltimore's most varied streetcar route was the No. 25 Falls Road–Mount Washington–Pimlico line. It included some Pittsburgh-style topography and a range of neighborhoods from nineteenth-century mill workers' communities to Victorian suburbs to the famous Pimlico racetrack. It also had history because pieces of the route were part of Leo Daft's pioneering 1885 electric line—this country's first successful street railway electrification. (Daft's tiny electric locomotives hauled horsecars between present-day Howard and West Twenty-fifth Streets and Hampden.)

Hampden • Woodberry • Mount Washington • Pimlico • Cheswolde

Stony Run valley is a woodsy, scenic barrier separating Hampden from downtown Baltimore. In the mid-1890s, the one-time City & Suburban Railway spanned it with this spectacular structure reaching from the end of Huntington Avenue to Thirty-third and Keswick in Hampden. Below was the "Ma & Pa" Railroad's single track. The trestle is in its last month of service in April 1949 as an inbound Peter Witt car rumbles across. R. W. JANSSEN

ABOVE | A picturesque appendage of the Falls Road line was the Union Avenue jerkwater, a short shuttle between Thirty-sixth Street and Roland Avenue in Hampden and the cluster of late-nineteenth-century textile mills and foundries in the Jones Falls valley at Woodberry. Semiconvertible No. 5765 drops down steep, narrow Union Avenue into the valley, passing old stone mill workers' houses. This anachronism lasted until April 1949, the date of this photo. G. J. VOITH   **ABOVE RIGHT** | The gateway to Mount Washington was the 1927 Kelly Avenue viaduct spanning the Jones Falls and the Pennsylvania Railroad's Northern Central line. Outbound PCC No. 7018 crosses in April 1949. The concrete viaduct replaced a spindly 1897 steel streetcar trestle and a grade-level street crossing a short distance to the north. G. J. VOITH   **MIDDLE RIGHT** | On April 24, 1949, the No. 25 line south of Kelly Avenue

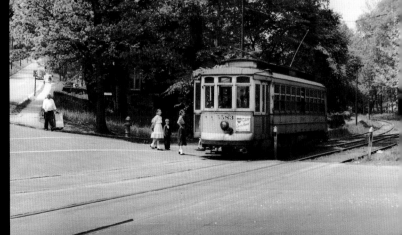

in Mount Washington was switched to bus, leaving the rail line from Mount Washington to Pimlico isolated from its downtown entry. This loop and bus transfer terminal was built at Kelly and Sulgrave Avenues as a temporary expedient until buses could be used over the full route, a service that finally began on September 14, 1950. In this May 1950 scene, a Peter Witt car from Belvedere and Pimlico is meeting a Falls Road bus, while in the left background a semiconvertible is pulling out on its way to Cheswolde.

E. S. MILLER   **BELOW RIGHT** | One reason for the two-step bus conversion of the No. 25 line was the difficulty in operating buses through the Western Run valley west of Mount Washington's commercial area. Roads following the route were primitive, and the

the valley and pauses at Kelly and Sulgrave Avenues in "downtown" Mount Washington. In May 1950, Kelly Avenue was little more than a path alongside the tracks. When the car line was abandoned, the avenue was completely rebuilt. E. S. MILLER

ABOVE | Baltimore seemed to abound in rustic pockets within the city, and Kelly Avenue was one. Here its primitive state is evident as No. 5583 eases downhill near Poplin Street, passing a tiny community built alongside Western Run. When the car line died, the community died, too, for it was almost entirely obliterated by the widening and repaving project. E. S. MILLER

ABOVE RIGHT | A Peter Witt car from Pimlico eases downhill toward Kelly Avenue

ABOVE | Onward to Pimlico. Outbound Peter Witt No. 6022 has finally surmounted the long, curving grades of Kelly Avenue and Cross Country Boulevard and rests at Cross Country and Ken Oak. This section was developed as Mount Washington Heights in the early 1900s; the car line itself was opened in 1897. E. S. MILLER **ABOVE RIGHT** | Lucky streetcar riders could catch snatches of the races at Pimlico as cars followed Pimlico Road along the east side of the track, seen to the left in this May 1950 photo. Peter Witt car No. 6021 is near Rogers Avenue on its way to Belvedere. E. S. MILLER **MIDDLE RIGHT** | The Cheswolde, or "Key Avenue," branch looked much like a turn-of-the-century rural trolley but actually was originally part of a main line from Baltimore to Pikesville, Reisterstown, and Glyndon. Built by the Baltimore & Northern in 1897, it followed Cross Country Boulevard and

Greenspring Avenue as far as present Cheswolde Road, where it turned west and headed directly across country to Reisterstown Road at Pikesville. With no nearby roads at the time, the section between Cheswolde and Pikesville was barely populated, and in 1923 the line was cut back to Cheswolde. This scene was typical: No. 5583 is outbound alongside a very rural Cross Country Boulevard west of Arlington Junction (Kelly Avenue) in 1950. E. S. MILLER

BELOW RIGHT | Cheswolde cars followed the side of Greenspring Avenue to their terminal at Cheswolde Road, originally called Key Avenue—hence, the name of the branch. Nearing the end of its run, this car is alongside Greenspring at Taney Road. E. S. MILLER

The thriving collection of car lines in north Baltimore disappeared in the late 1940s as most of the major north-south streets—notably Calvert, Saint Paul, and Charles Streets and Maryland Avenue—were converted to one-way operation to speed auto traffic. The bus conversion was a mixed blessing: the routes to Hampden, Roland Park, and Guilford had varying amounts of private track and could move fast, but south of North Avenue their downtown routings tended to be tortuous and time consuming.

Outbound on the No. 25 line, PCC car No. 7020 is on Maryland Avenue at Lafayette, bound for Mount Washington and the Belvedere loop in 1946. Poking out of Lafayette at the left is a No. 30 Fremont Avenue line car that has just made its terminal loop and will work its way west on North Avenue. G. J. VOITH

| The No. 1 Gilmor Street line cars, such as this semiconvertible southbound on Saint Paul at Twenty-first Street, came out of town on Guilford Avenue (including a stretch on the elevated) to North Avenue, then headed west on North Avenue to Saint Paul, north on Saint Paul to Twenty-fifth, and then east again to a terminal at Greenmount Avenue. But Baltimore was slower then, as this 1947 scene of the "placid rows" so strongly suggests. G. J. VOITH

ABOVE RIGHT | Typical of the convoluted downtown routings was the one taken by this No. 29 Roland Park car, shown turning north onto Saint Paul Street from North Avenue near Polytechnic High School in 1947. To reach this point, it has traveled on Calvert Street, Read Street, Charles Street, and North Avenue; it will proceed north on Saint Paul to University Parkway, where it will head west. G. J. VOITH   **BELOW RIGHT** | Skirting the north side of the Johns Hopkins University campus, an inbound PCC car from Roland Park is rolling eastward on University Parkway near Charles Street. By this time car riders were buying shiny new autos like the 1946 Buick whizzing past on the right or the more lowly Plymouth parked on the left. G. J. VOITH

ABOVE RIGHT | In 1916, a unique branch line was built to serve the newly developed and highly exclusive Guilford area. Continuing the Saint Paul Street trackage, the new line followed the northern extension of Saint Paul Street past wide lawns and large homes to Bedford Square at Charles Street. Its double track was split on the two sides of the roadway. Trimmed hedges separated the track from the street, making the car line not only unobtrusive but almost invisible. Typical was this scene just east of Bedford Square on Saint Paul Street in 1947. The viewer looks northwest toward Charles Street. G. J. VOITH

BELOW RIGHT | The Guilford branch ended at Bedford Square, where the divided tracks came together at an attractive waiting station. Here the viewer looks southeast on Saint Paul Street as No. 5591 arrives, while another semiconvertible pulls away. G. J. VOITH

Roland Park was a pioneering planned upper-class suburban community, first laid out in the early 1890s and landscaped by Frederick Law Olmsted's firm. Part of the project was a "rapid transit" line downtown, following picturesque private track on Roland Avenue and, later, University Parkway.

ABOVE | Roland Park's commercial center was this aesthetic building at Roland and Upland Avenues. Built in 1896, it was the country's first integrated suburban shopping center—the great-grandfather of today's massive malls. The No. 29 line terminated here; in earlier days, so did the No. 24 Lakeside shuttle and the No. 10 line from Hampden via lower Roland Avenue, making the shopping center a busy transfer point. In this July 1946 scene, No. 5797, with a railfan excursion, has followed a regular No. 29 PCC car into the terminal station. R. S. CROCKETT **ABOVE RIGHT** | In this scene looking southeast on Roland Avenue at University Parkway, a No. 29 Roland Park car has just entered Roland Avenue after its scenic climb along a tree-lined private reservation on University Parkway. At the right rear, a No. 24 Lakeside shuttle run makes ready to follow it out Roland Avenue.

MIDDLE RIGHT | Directly behind the Roland Park terminal loop was this carhouse, which, like the shopping center, dated to 1896. Its original aesthetic stone façade with multiple arched doorways proved unable to accommodate larger equipment, and it was altered to this form in 1908. R. S. CROCKETT **BELOW RIGHT** | From the shopping center, car tracks continued north on Roland Avenue to its end at Lake Avenue and then dipped into the woods for a short but scenic run to Lakeside Park, a one-time picnic spot overlooking Lake Roland. This lightly populated upper Roland Avenue section was served by a separate line, No. 24, running from the Roland Park water tower at Roland Avenue and University Parkway to Lakeside Park. Here the viewer looks south on Roland

ABOVE │ In June 1946, No. 5388 negotiates the rustic Lakeside loop, which was located on a hill above Lake Roland. The park had a brief life; it was created in 1897 and closed before 1909. But its two-story pavilion managed to survive and outlast the streetcar line. R. S. CROCKETT   ABOVE RIGHT │ A little of York Road's country feeling remained in this scene south of Stevenson Lane on a late afternoon in October 1963, as southbound No. 7077 paused for a pickup. Within a month both the leaves and the car line would be gone. H. H. HARWOOD, JR.   MIDDLE RIGHT │ Again, York Road was deceptively quiet on a Sunday morning as inbound No. 7316 passed the old Sheppard Pratt Hospital entrance opposite Worcester Road in May 1951. Until 1930, the car tracks through this area were set in a private reservation on the east (right) side of the old two-lane turnpike. The waiting station on the left survived to serve

bus riders but has since vanished. E. S. MILLER   BELOW RIGHT | Before 1958, Towson had both railroad and streetcar service. The two crossed at York Road and Susquehanna Avenue, shown here in May 1951 as Irvington-bound No. 7334 passed under the "Ma & Pa" bridge. A block to the left, the frame railroad station was still a stopping place for the leisurely passenger run to such points as Glenarm, Bel Air, Delta, and York. The stone bridge abutments continue to be York Road landmarks, although the frame house behind has been replaced by the Towson branch of the Baltimore County Public Library. E. S. MILLER

Baltimore's heaviest car line, the No. 8 route to Towson, followed York Road. In the days before Interstate 83, York Road was also the primary motor route to York and Harrisburg, Pennsylvania, and points north. In its early days, the No. 8 line had been a long country trolley ride to a sleepy county seat, passing large estates and summer homes along the way. But its charm diminished as the York Road corridor commercialized and suburbanized and as motor vehicles multiplied. Things were relatively quiet on this Sunday morning in mid-May 1951, when a southbound car passed the landmark Senator Theater, then merely one of many neighborhood movie houses. Unlike many scenes in this book, the foreground buildings have changed little, although the bucolic background has mostly disappeared in favor of more commercial development and the extension of Northern Parkway. E. S. MILLER

ABOVE | Towson nocturne: No. 7106 lays over on an evening in October 1963. H. H. HARWOOD, JR.

ABOVE RIGHT | Towson streetcars ended their runs on Washington Avenue in front of the 1854 county courthouse (out of the photo to the left). The scene on the opposite side of Washington Avenue in 1959, when No. 7359 waited to start its long journey to Catonsville, was much different from that of today. The block is now almost wholly occupied by new office buildings. A. W. MAGINNIS

MIDDLE RIGHT | From the other side of the street, the courthouse dome pokes up between the leaves at the left. Between 1912 and 1923, the hapless, battery-powered Towson & Cockeysville Electric Railway came into the square from the far left of this 1963 photo. H. H. HARWOOD, JR.

BELOW RIGHT | Leaving their Towson courthouse terminal, No. 8 cars returned to York Road on Chesapeake Avenue. Towson's small town atmosphere was clearly still strong in 1951 when Ed Miller caught PCC car No. 7380 making its pickup stop before turning south. Today, newer buildings occupy both sides of the corner—most notably the Towson Commons complex, which replaced

Baltimore had no major league sports teams in the 1940s, but crowds still came to the old Baltimore Stadium, Memorial Stadium's predecessor on the Thirty-third Street site. And many came by special streetcars, which reached the stadium over a spur running on Loch Raven Road (as it was then called) from Gorsuch Avenue to Thirty-sixth Street. Since the stadium branch had no regular service, the cars simply were stored in long lines along Loch Raven until the game ended.

It is early summer in 1946, and the original minor league Orioles are playing, having been burned out of their own ballpark on Greenmount Avenue in Waverly two years before. Prophetically, the lead car in this game lineup will go to Camden Station (the railroad terminal); for present-day baseball games, light rail cars go to Camden Yards. R. S. CROCKETT

This lineup on Loch Raven Road is awaiting the end of the 1944 Army-Navy football game. G. J. VOITH

A workaday northeast Baltimore trunk line was the No. 19 Harford Road route, which came out of downtown via Aisquith Street and Central Avenue. This outbound Parkville car is snaking through the S curve at the Central Avenue–Harford Road intersection in May 1950. E. S. MILLER

The long, straight routes out Harford and Belair Roads originally were semirural, side-of-the-road lines serving small turnpike settlements. But rapid residential and commercial development along these arteries swallowed up the open fields, and street-widening projects displaced the one-time private track. By the 1940s, both were rather ordinary streetcar lines noted more for traffic volume than scenery.

ABOVE | Another pretty interlude on Harford Road was the dip through the park in the Herring Run valley. Once called Hall's Springs, after a still active spring to the left of the photo, this was the terminal of a horsecar line that struggled out Harford Road in the early 1870s. The one-time summer resort was much changed in 1952, as an inbound PCC dropped into the valley at Walther Avenue. E. S. MILLER

ABOVE | The No. 15 Belair Road line was another heavy passenger carrier. Wearing the short-lived yellow-green and light gray "fruit salad" colors of BTC's new owner, National City Lines, PCC No. 7038 heads into town on Belair Road at Sinclair Lane in July 1951. Behind is the B&O Railroad's Baltimore Belt Line, part of its main line to Philadelphia. Note the railroad's little wood shelter station on the right, behind the used car lot. Commuter locals once stopped here and took passengers to Mount Royal and Camden Stations. E. S. MILLER ABOVE RIGHT | Harford Road cars ended their runs at the Parkville loop, just north of Taylor Avenue. Between 1904 and 1936, a single-track rural jerkwater line extended beyond here to Carney, at Harford and Joppa Roads. Here, in April 1952, the Carney trolley's diesel replacement has picked up five passengers from the No. 19 line PCC cars at the left and soon

will head north. Harford Road trolleys ran until June 16, 1956. E. S. MILLER **MIDDLE RIGHT** | The Belair Road line was not particularly scenic, but the east side of Clifton Park formed a pretty summer backdrop for yellow-orange Peter Witt No. 6032 running inbound near Cliftmont in July 1951. E. S. MILLER **BELOW RIGHT** | Belair Road's rolling topography added a little life to an otherwise bland trip on the No. 15 line. Here an inbound PCC car comes up out of the Herring Run valley at Pelham Avenue in mid-1951. Like the York Road and Harford Road routes, this line was originally laid on a roadside private right of way but was relocated during street-widening and repaving projects in 1928 and again in 1938. But at least the streetcar line survived, for Belair Road was

East Baltimore once was crisscrossed with lines on such streets as Caroline, Preston, Federal, Orleans, and Patterson Park Avenue; most were replaced by buses or trackless trolleys before World War II. One postwar survivor was the route out Monument Street to Kresson, in the Orangeville section of town, then a part of the No. 6 Curtis Bay line. Basically another "row-house line," its eastern extremity (built in stages between 1910 and 1917) was laid on a private right of way along the north side of Monument Street through a rather ponde-

Cars of the No. 18 and No. 34 lines wound through the working community of Canton south of Eastern Avenue. Semiconvertible No. 5772, working the No. 18 Pennsylvania Avenue–Canton line, weaves its way on Essex Street at the Fait Street–Montford Avenue intersection in 1949. E. S. MILLER

ABOVE | More charming were the streets in Canton, particularly when decorated by the creaky semiconvertibles. This No. 18 line car is at South Kenwood Avenue at Elliott Street in November 1949. E. S. MILLER **ABOVE RIGHT** | Outer Monument Street was not an uplifting sight, but east of Highland Avenue cars of the No. 6 line could at least skim along on their own right of way on the north side of the street. In this view looking east, an inbound PCC car rolls along the trash-littered roadside track heading for its faraway terminal at Curtis Bay. A large brickyard was at the left. After the car line died, Monument Street was widened over the area of the car tracks and used auto dealers took over some of the brickyard space. G. J. VOITH

MIDDLE RIGHT | Near their Kresson Street terminal, Monument Street cars passed under two Pennsylvania Railroad bridges—in the foreground, the branch to Canton and, in the rear, the high-speed main line to Philadelphia and New York, now Amtrak's Northeast Corridor route. Peter Witt No. 6036 is returning west in 1946; two-lane Monument Street is at the right. G. J. VOITH

BELOW RIGHT | A No. 34 "Third Street"–Canton car, No. 5727, changes ends at its stub terminal on Highland Avenue near

─── This track continued south, eventually ending at Clinton Street and Holabird Avenue near the Canton water

Another Canton-area route was the No. 20 Point Breeze line, which served a more industrialized area along Oldham Street and Broening Highway. In earlier years it was famous as the route to Riverview Park at Colgate Creek; later it carried hordes of workers to the Western Electric plant that had been built on the old park site. En route, it worked its way through a maze of B&O and Pennsylvania Railroad industrial tracks, such as this crossing of the Pennsy's Dundalk–Sparrows Point branch at Sixteenth

MIDDLE RIGHT | A major landmark in Highlandtown, or "Hollandtown" in the local patois, is the elaborate series of underpasses carrying Eastern Avenue under the one-time Pennsylvania Railroad's industrial branches to Canton and Fells Point. This 1947 view looks east as a two-car train of "red rockets" from Sparrows Point whines through the long dip. In the distance behind, a Pennsy steam switcher shuffles a cut of cars on the branch to Canton and Fells Point. This line, opened in 1838, terminated at President Street Station and was the original route between Baltimore and Philadelphia. G. J. VOITH   **BELOW RIGHT** | Looking west toward Haven Street from beneath the railroad overpass, the camera catches a pair of outbound "rockets." G. J. VOITH

Other East Baltimore lines were sideshows to the main event: the heavy, blue-collar No. 26 Sparrows Point line serving the big Bethlehem Steel mill and shipyard complex at "The Point" and its bedroom communities of Dundalk and Turners Station. The mill was first developed in this remote southeastern corner of Baltimore's harbor in the late 1880s. In the mill's early days, workers either lived in the company town of Sparrows Point or rode to work on Pennsylvania Railroad local trains. The streetcar line across the flatlands and the wide Bear Creek estuary was completed in 1903; three years later it was extended beyond to Fort Howard and Bay Shore Park.

Dundalk • Turners Station • Sparrows Point • Bay Shore Park • Fort Howard

For much of its length beyond Highlandtown, the Sparrows Point route followed Dundalk Avenue and its extension, Main Street in Turners Station. It was straight, fast, and heavy-duty—close, in fact, to modern light rail lines. Railroad-style automatic block signals protected the line, and "crossbuck" warning signs stood at street crossings. Here a single "red rocket" passes Fait Avenue north of Dundalk, now the site of the I-95 interchange, in May 1950. E. S. MILLER

ABOVE | "Test Drive the '50 Ford," says the sign on the car on the left, and many wartime trolley riders were buying. Also accompanying "red rocket" No. 5867 at Dundalk Avenue and Bayship Road are a classic Good Humor truck and an old bus made into a mobile lunchroom. The streetcar is bound for Sparrows Point in July 1950. E. S. MILLER

ABOVE RIGHT | The community of Dundalk was farmland before World War I, when Bethlehem Steel and the U.S. Shipping Board began development of a carefully planned suburb for the booming "Point." Dundalk was a pragmatic spot; its center was located at Center Place, whose center in turn was the trolley station. Single semiconvertible No. 5847 is loading for Highlandtown and down-

LEFT | No. 5787 is at Maryland Avenue and No. 5785 is at the Riverside power plant in July 1950. Note the "crossbuck" guarding the street crossing and the railroad-style block signals. The signals, spaced at regular intervals, protected against rear-end collisions on this fast track. E. S. MILLER

ABOVE | Sparrows Point lurks in its perpetual haze in the distance as cars go "out to sea" across Bear Creek. E. S. MILLER

| One of the most spectacular streetcar structures anywhere was the long trestle spanning Bear Creek, a wide bay estuary separating Sparrows Point from the "mainland." Highways into "The Point" dodged around to the north to cross where the creek was narrower; the car line struck straight across from Sollers Point. In August 1950, PCC cars were strongly in evidence as a Sparrows Point semiconvertible meets its replacement at the Sollers Point end. E. S. MILLER | The swing span is open, probably for one of the many pleasure boats based in the various inlets in the area, in this April 1956 view from the front of a Sparrows Point car. The signal at the right was a variation of the "smashboard" design; not only was it an unmistakable stop signal, but should a car run through it, the blade would be broken and thus provide positive evidence that the motorman disobeyed the signal. H. H. HARWOOD, JR.

MIDDLE RIGHT | On the Sparrows Point side, both the scenery and the atmosphere suddenly degenerated. Car riders were immediately plunged into Bethlehem Steel's sprawling mill and shipyard complex, one of the largest in the United States. A short distance beyond the trestle, a branch swung south to a loop at the shipyard entrance. In September 1953 the viewer looks east on the main line into downtown Sparrows Point. The shipyard spur, opened in 1919, branches to the right. E. S. MILLER

BELOW RIGHT | Skirting steel mills and yards of the Patapsco & Back Rivers Railroad, Bethlehem's switching subsidiary, PCC car No. 7057 is on its way to the Sparrows Point terminal in April 1958, four months before service ended. The car is swinging onto a mill road that carried streetcar and motor traffic over the P&BR tracks on a steel viaduct. F. W. SCHNEIDER III

ABOVE | Sparrows Point terminal, at the east end of town, was a busy spot, especially in the summer when crowds heading for Bay Shore Park transferred here to the shuttles running east to the park on the Chesapeake Bay. Shuttle cars for Fort Howard also connected here. This view looks east in 1945 as a two-car "red rocket" train loads for Baltimore. In the distance is the right of way to Bay Shore and Fort Howard. One of the two tracks usually was used for car storage. G. J. VOITH ABOVE RIGHT | A railfan excursion is rounding the loop terminal on the shipyard branch in 1953. R. W. JANSSEN MIDDLE RIGHT | Nested in the midst of the mills was the town of Sparrows Point, the company-built community laid out in the days when the "Point" was remote from

into the town from the north on Fourth Street. In November 1950, semiconvertible No. 5857 trundled down D Street as track and paving work was under way. In Sparrows Point's stratified society, D Street houses were reserved for middle- and lower-level managers. Now the town is gone. What one newspaper once called "the cleanest and greenest steel mill town in the USA" was bulldozed for mill expansion in 1973. E. S. MILLER   BELOW RIGHT | The scene was a good bit more barren by 1956. Once a wooded haven beyond the mills, the Sparrows Point terminal looked like this in its later days. In 1950, a loop had been built for the PCCs. The Bay Shore branch died after the 1946 summer season, and Fort Howard service ended in 1953. Afterward, mill expansion swallowed up

ABOVE | Built in 1906, the Bay Shore–Fort Howard extension seemed to be constantly hopping creeks. Originally built as a large loop circling the peninsula along the Chesapeake Bay, the line was cut into two stub-end branches when a 1933 hurricane removed the trestle over Shallow Creek in the loop's center. This trestle carried the two lines over Jones Creek, a short distance east of Sparrows Point. A Fort Howard shuttle car crosses in October 1946. R. S. CROCKETT

ABOVE RIGHT | The single-track branches to Bay Shore Park and Fort Howard diverged at a spot in the woods called North Point

MIDDLE RIGHT | The Chesapeake Bay lapped at the streetcar track as the cars approached Bay Shore Park. G. J. VOITH

BELOW RIGHT | A one-time United Railways & Electric Company trolley park—and one of Baltimore's most popular—Bay Shore opened in 1906. Soon after, crowds came in by the trainloads for swimming in the bay, picnicking, and riding the park rides, which eventually included a roller coaster. At the end of the line in 1945, semiconvertible No. 5842 waits to return to Sparrows Point.

Bay Shore cars loaded under a large trainshed next to the roller coaster in the park's center. In this view the shed can be seen in the center; the track in the foreground—once part of the through line from Fort Howard—was used as a storage spur. This may have been the only place in the United States where a streetcar track passed under a roller coaster. Bethlehem Steel bought the park property in 1947 and demolished its buildings but never used the site. Oddly, however, the streetcar terminal shed survived and has since been restored as part of a new park. C. W. HOUSER, SR.; BALTIMORE CHAPTER NRHS COLLECTION

The areas south of downtown Baltimore were, and still are, working communities oriented to the docks and industries that developed around the south shore of the harbor: shipbuilding, chemicals, petroleum, scrap, and sugar refining. And like their surroundings, the streetcar lines serving them were unglamorous but offered much vitality and interest.

Locust Point • Westport • Brooklyn • Fairfield • Curtis Bay

The No. 2 Fort Avenue line crossed Baltimore's other "Point"—Locust Point, a beehive of B&O and Western Maryland railroad piers, grain elevators, shipyards, and port-related industry. But it also ended at Baltimore's premier tourist attraction, Fort McHenry. Here No. 5201 poses outside the fort gate during a 1947 railfan excursion. Regular cars ended their runs on a loop at the right. G. J. VOITH

ABOVE | Passengers inside PCC No. 7108 brace themselves for the jolts as the inbound Westport car crosses the four-track B&O Railroad main line at Ridgely Street in 1946. In the days before the Russell Street bridge, traffic often jammed up at this active crossing, waiting for the parades of passenger trains, freights, light engines, and Baltimore & Annapolis interurban trains. Between 1935 and 1950, the B&A used these tracks from Camden Station to its own line near Westport. G. J. VOITH

ABOVE RIGHT | Inbound semiconvertible No. 5648 picks up a hitchhiker on Annapolis Road in Westport. Just ahead are the Western Maryland Railway crossing and then the B&O's Curtis Bay branch crossing. G. J. VOITH

BELOW RIGHT | Nearing its terminal at Waterview Avenue, a PCC car rolls up the hill on Annapolis Road in 1947. Streetcars first came to Westport in 1896. By the early 1900s, the little community on the Patapsco's Middle Branch also had two steam railroads (the B&O and the Western Maryland) and two electric interurban lines—the Washington, Baltimore & Annapolis and the Annapolis Short Line. G. J. VOITH

ABOVE | Mainstays of the Curtis Bay–Fairfield line, the austere looking Peter Witts seemed to fit their bleak surroundings. This one is returning downtown from the Fairfield branch, which formed a large loop serving the shipyards and refineries on the East Brooklyn–Fairfield peninsula. The viewer looks east from the B&O Railroad overpass at Patapsco Avenue and Shell Road on the last day of service, March 20, 1948. R. S. CROCKETT

ABOVE RIGHT | Westport still had two electric lines when this photo was taken in 1947. Emerging from the one-time Washington, Baltimore & Annapolis's 1908 Westport tunnel is Baltimore & Annapolis No. 205, bound for Annapolis. Directly above, a Baltimore Transit Company PCC car lays over at the Westport terminal loop at Annapolis Road and Waterview Avenue. This scene has been

obliterated by the Baltimore-Washington Parkway, which occupies the area immediately to the right of the tunnel. Oddly, however, the electric railway tunnel remains intact, although well disguised. R. S. CROCKETT

BELOW RIGHT   The streetcar gateway to Brooklyn and Curtis Bay was the impressive 1917 Hanover Street bridge, which in the mid-1940s carried a constant parade of No. 6 streetcars to and from the heavy industries, shipyards, and coal piers at Curtis Bay and Fairfield. Peter Witt No. 6074 is on its way south in 1947. G. J. VOITH

ABOVE | Inbound Peter Witt No. 6051 comes around the west side of the Fairfield loop in 1947. This branch was a World War I project, opened in 1918. In 1941 the original loop was extended to the northeast to reach closer to the wartime Bethlehem Steel shipyard and the Maryland Shipbuilding & Drydock facilities. G. J. VOITH

ABOVE RIGHT | The line into Curtis Bay was perhaps even grimmer than the Fairfield branch, as this view along Curtis Avenue attests. G. J. VOITH

BELOW RIGHT | The Curtis Bay loop terminal was in this no man's land south of Aspen Street and Curtis Avenue, seen just as service was ending. R. S. CROCKETT

Like that of most streetcar systems, Baltimore Transit's work equipment was a melange of hand-me-downs, rebuilds, and utilitarian built-for-the-purpose cars.

Test car No. 3550, used to test rail bonding, fell into the hand-me-down category. A 1904 Brill product, it obviously started life as a passenger car. Here it suns itself at the Govans loop on November 2, 1963, the last day of streetcar service. The Baltimore Streetcar Museum saved it and has since impressively restored it to its 1924 passenger configuration and appearance, including its original number and brilliant orange and cream colors. H. H. HARWOOD, JR.

A genuine high-wire act: line car No. 3503 and its nerveless crew work on the Guilford "el" overhead. By the time this July 1947 photo was taken, motor trucks handled most wire maintenance work, but private rights of way and locations such as this obviously required a railcar. Note the grain elevator in the background, later a casualty of the Jones Falls Expressway extension. R. W. JANSSEN

In the streetcar era the transit company was responsible for keeping its tracks clear during snowstorms—considerably easing the work of the city's street crews. With its brooms whirling, sweeper No. 3239 fights its way east on North Avenue at Maryland Avenue during a heavy (for Baltimore) fall in December 1948. R. W. JANSSEN

Crane No. 3736 blocks a line of trackless trolleys at Howard and Lombard Streets in October 1947 as it removes rail from a crossing abandoned more than seven years earlier. Note the "WB&A" sign at the left, advertising a restaurant in the old Washington, Baltimore & Annapolis terminal station. R. W. JANSSEN

Survivors

Three 1860s-era pavilion stations of the short-lived City Park Railway (the "dummy line") survive inside Druid Hill Park. Two have been relocated from other sites, but the little railroad's terminal, the flamboyant Council Grove pavilion at the zoo entrance, stands on its original site. Here it is in its heyday, sometime in the 1870s. The equally flamboyant little self-propelled passenger car housed a small steam engine at one end. R. M. VOGEL COLLECTION

As this is written in 2002, it has been thirty-nine years since Baltimore's last true streetcar ran. Many lines have been gone for forty-five years or more. Their tracks long since have been covered with asphalt or removed completely. What was once grassy, open private track has been made into new streets, swallowed up by street widening, or simply left to revert to nature. Occasionally, you can spot bits of rail tentatively poking up through the pavement or perhaps make out telltale ridges running down the center of some street, marking thinly buried car line. But these are not much more than fleeting reminders, for the next street-surfacing project will submerge the rails again or finally extract them.

In some sections of the city, the entire environment in which the streetcars operated has been obliterated. Indeed, it is impossible to relate trolleys to the present Inner Harbor development or much else of Baltimore's downtown area. It is equally difficult to picture what is now the location of the Harford Road exit of the Beltway as a two-lane rural road with the single track of the Carney jerkwater running alongside it. Seemingly, the city's streetcar system has vanished completely and irrevocably.

Well, not entirely. The rails themselves may not be visible, although many still lurk under the street surface. But quite a few of the structures that served the cars and sheltered their passengers can be seen today—if you know where to look. In fact, Baltimore has what is probably the richest and most comprehensive collection of old street railway buildings anywhere. Still standing, and mostly in healthy condition, are three ornate stations of a Civil War—era steam dummy line, a horsecar terminal, four cable railway powerhouses, a large turn-of-the-century shop and powerhouse, and a miscellaneous assortment of carhouses and terminal buildings. In short, chapters from the entire history of the street railway in Baltimore still exist in brick, stone, and concrete.

The mandatory starting point for any street railway archeologist is at the southern end of Druid Hill Park. Here, in the space of only a few blocks, are relics representing the oldest, newest, and almost every phase in between. Almost from its earliest days, the park was a focus for car lines, and important terminal structures were regularly built there and then discarded for something newer. Happily, the area was bypassed by later development and also somehow escaped urban renewal, thus allowing much to remain amazingly intact.

Let us start with the oldest. Inside the park itself, near its southern entrance, is a gingerbready open pavilion, which has been recently relocated from its original site on the northwest corner of Druid Hill and Fulton Avenues. Probably built in the mid-1860s, this originally served as a station on the City Park Railway, a horsecar and later steam dummy line built in 1863 to connect the then-new park with North Avenue. As built, the little station was an exuberant Victorian-Oriental confection with a pagoda-like roof, once called the "Chinese station." The Park Railway was dismantled in 1879, but the station-pavilion continued in use, successively serving city horsecars, cable cars, electric trolleys, and buses. Its roof ornamentation long since has been stripped and subdued and the structure itself was reduced in size, but much is still original.

Elsewhere inside Druid Hill Park are two other one-time Park Railway stations, both of them even more fanciful. The large Council Grove pavilion opposite the zoo entrance was the line's northern terminal. After years of neglect and a close brush with destruction, it was restored in 1972 and now looks precisely as it did when the little steam cars loaded there. The "dummy line" also had a way station at what is now Auchentoroly Terrace and Orem Avenue. Smaller than the others but equally exotic, this Oriental-eclectic structure subsequently was picked up and moved to the hillside overlooking Druid Lake, where it lives on as the Latrobe pavilion.

Just outside the park and across the street from where the old "Chinese station" originally stood is the architectural opposite of the whimsical "dummy line" stations—the huge fortresslike former Park Terminal. Dominating the southwest corner of Druid Hill and Fulton Avenues, Park Terminal was built in 1909 to replace the mishmash of obsolete carhouses in the area and was the model of a modern, efficient streetcar storage and servicing facility. It also originally housed company offices. Closed in 1952 and now owned by the city, it has been remodeled slightly but is essentially unchanged. Look above its two main entrance doors, and you will see *Park Terminal* carved in stone over one

And Council Grove as it looks today, having been beautifully restored after a close brush with destruction. The scrollsaw woodwork on the roof was made new when the structure was restored, but it recreates the original embellishments, which had been removed many years earlier. H. H. HARWOOD, JR.

The Victorian-Moorish Latrobe pavilion was another "dummy line" station, originally located near Auchentoroly Terrace and Orem Avenue. It was later relocated to the east side of Druid Lake, where it still stands. H. H. HARWOOD, JR.

and *Waiting Room* over the other. Immediately south of Park Terminal, and part of the same complex, is the brick 1890s-era terminal of the Central Railway, which was later used as a substation building.

Several blocks south of Park Terminal, at the southwest corner of Druid Hill Avenue and Retreat Street, is a ponderously ornate stone hulk that is the country's finest survivor of the short-lived cable railway era. Built by the Baltimore Traction Company in 1890–91, the massive structure was a combination carhouse and powerhouse. Inside its north end, two Corliss stationary steam engines operated the understreet cables that moved the cars on the northern half of Baltimore Traction's Druid Hill–Patterson Park route. After abandonment of the cable system in 1896, the building briefly was used as an electric carhouse. Now, with most of its windows and doors bricked up, it's a warehouse. But there's no question about its heritage—the name *Baltimore Traction* clearly stands out in stone over the warehouse doors.

Less than two blocks to the east of this was once Baltimore's most elegant late Victorian carhouse. It is there still but only as a semblance of what it once was. The present single-story warehouse in the block bounded by Madison Avenue, Cloverdale, and McCulloh Streets was originally the carhouse and general office of the Baltimore City Passenger Railway—still another of the fiercely independent companies in the days before the "United." Its northeast corner originally incorporated a three-story office building alive with spires, gables, and ornamental ironwork. The carhouse was sold in 1947, and the two top stories were cut down soon after. A few original iron filigrees still hang on from the glory days.

Other bits of street railway history can be found in all corners of the city. Three other cable powerhouses survive—more, in fact, than anywhere else in the country. Two of them are close neighbors in East Baltimore —one on the north side of Pratt Street west of Central Avenue (built by Baltimore Traction in 1890–91) and the other on the north side of East Baltimore Street at Aisquith (dating to 1892–93). The cable system turned out to be an expensive mistake, and the lines were electrified in the late 1890s; neither building served its original purpose for more than five years. The Pratt Street structure is now a city-owned garage; the East Baltimore Street building has had a variety of uses, recently

Quite elegant by horsecar terminal standards, the one-time Peoples Railway carhouse and general office on the east side of Druid Hill Avenue at Retreat Street was built in 1885. Note the words *Peoples . . . Co.* over the front door; the "Railway" part has been covered by the modern sign mounts. The carhouse was obsolete within ten years of its completion and was never used for electric streetcars. Unfortunately, this rare relic was lost to a fire soon after this 1984 photo was made. H. H. HARWOOD, JR.

The forbidding Romanesque Druid Hill Avenue cable powerhouse and carhouse dates to 1891. The viewer faces north on Druid Hill Avenue toward Retreat Street. The nearest section was the carhouse; the turreted adjoining portion housed the stationary steam engines that propelled the understreet cables. The name *Baltimore Traction Co.* can be seen carved in stone over the one-time carhouse doors. This complex is directly across the street from the now-departed Peoples Railway horsecar terminal.

H. H. HARWOOD, JR.

Park Terminal, as seen from the Druid Hill Avenue side in 1983. The passenger waiting station was at the northeast corner, seen at the far right in this photograph. UR&E division operating offices originally were on the second floor. H. H. HARWOOD, JR.

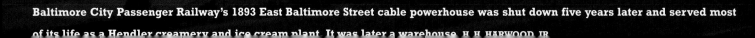

Baltimore City Passenger Railway's 1893 East Baltimore Street cable powerhouse was shut down five years later and served most of its life as a Hendler creamery and ice cream plant. It was later a warehouse. H. H. HARWOOD, JR.

as an ice cream plant. The third surviving powerhouse is probably more familiar. Located on the east side of Charles Street north of Lanvale, the one-time Baltimore City Passenger Railway powerhouse and its adjoining carhouse are both part of the expanded Charles Theater. Sharp-eyed observers can spot *BCPRCo.,* for Baltimore City Passenger Railway Company, carved in stone on the cornice of the former powerhouse portion of the theater.

When Baltimore's gaggle of independent streetcar companies was consolidated into the United Railways & Electric Company in 1899, one of the UR&E's first major projects was a single, centralized shop to handle all heavy repairs and car rebuilding for the system. It picked a vast piece of property on Washington Boulevard opposite Carroll Park and started from scratch with a first-class facility. E. Francis Baldwin, a creative and certainly versatile Baltimore architect, was hired to design the shop. The prolific Baldwin had produced an incredible variety of commercial, ecclesiastical, and industrial buildings in Baltimore and elsewhere—including such past and present landmarks as the Mount Royal Station, Maryland Club, Fidelity Building, Old City College, and B&O Railroad Museum roundhouse, as well as the now-departed Rennert Hotel and Saint Mary's Seminary. Baldwin also had built a string of very attractive stations along the B&O Railroad system. Carroll Park shop was completed in 1901 and consisted of two immense monitor-roofed buildings incorporating four bays each, plus extensive outside storage trackage. Baldwin built well at Carroll Park; the shop survived the transition from streetcar to bus and is still the operating and maintenance heart of the MTA bus system, the "United's" lineal descendent.

The UR&E also located new carhouses strategically around the city to supplement and replace the motley assortment of obsolete structures inherited from its predecessors. Similar in style to Park Terminal, they were solid, stolid structures built of brick and concrete; complete with parapets and battlements, they looked like they could withstand anything. And withstand they did. All but one were closed as the trolleys were phased out, but most of the buildings were saved by their efficient design and low-maintenance construction. Still standing at the beginning of the twenty-first century (in addition to Park Terminal) are carhouses on Edmondson Avenue at Poplar Grove, York Road near

Its close neighbor, the 1891 Baltimore Traction East Pratt Street powerhouse, is a contrast in architectural style and state of preservation. The viewer looks north from Central Avenue in 1975. H. H. HARWOOD, JR.

Recycled and altered several times over, the present Charles Theatre complex was built by the Baltimore City Passenger Railway in 1893. The section nearest the camera was the cable railway powerhouse and housed steam engines for the company's north-side cable route, which ran to Twenty-fifth and Saint Paul via Calvert, Charles, and Saint Paul Streets. It first became a theater in 1939. Beyond the powerhouse is the former carhouse. When the cable operation was converted to electric power, this structure proved unsuited for housing the new streetcars and was successively a bus garage, a dance hall, and finally a part of the adjoining five-screen movie theater. The exterior of the complex has been restored to look as close to the original as was practical. H. H. HARWOOD, JR.

Northway, North Avenue west of Gay Street, and Lombard Street at Haven in Highlandtown. All were built in the 1907-8 period. Incidentally, most if not all of these also came off the drawing board of Francis Baldwin.

Several other older carhouses also serve varying uses—one on West Twenty-fifth Street at Howard, now an auto dealer, and one on West Baltimore Street at Smallwood. Another is located at Preston and Potomac Streets, and a one-time horsecar terminal and stable stands on Thames Street in Fells Point. In Owings Mills, the Baltimore & Northern's 1897 carhouse and powerhouse still survive on Reisterstown Road near the former Western Maryland Railway crossing.

Although usually not thought of as a street railway structure, the hulking Pier 4 powerhouse in the Inner Harbor complex was another United Railways & Electric project. Originally built to centralize streetcar power generation, the big, coal-fired plant was begun about 1900 but completed in at least three different sections at varying later times. It was situated to receive coal from barges via an elaborate conveyor system, since removed, and was also reached by railroad sidings from the B&O's switching line along Pratt Street. Idle for many years, the old plant seemed to defy any adaptive use and came perilously close to demolition. But after several false starts, it finally found a secure—if somewhat gaudy—life as a multiuse entertainment, restaurant, and retail center.

Also scattered around the city are several small, former streetcar waiting stations, such as the attractive little structure at Bedford Square (Charles and St. Paul Streets). Still in fine condition, this one was built about 1916 as the terminal of the Guilford car line; later, as Homeland was built up to the north, it also served as the transfer point for a feeder bus line. Somewhat lesser trolley stations stand at University Parkway east of Roland Avenue and at Catonsville Junction (Edmondson Avenue east of Dutton). In Mount Washington, a relatively elaborate 1897 trolley station survives at the northeast corner of Kelly Avenue and Sulgrave. Extensively altered, it is now an animal hospital. And on Belair Road north of Northern Parkway, the old Overlea streetcar terminal station underwent a death and rebirth. After surviving the end of streetcar service by thirty-seven years, it finally reached

York Road carhouse, then and now. Typical of the "modern" 1907 and 1908 United Railways & Electric carhouses, the big York Road structure was one of the last two active survivors. Seen in this October 1963 view (this page) just before the end of service, it later (next page) saw service as a lumber and hardware store and then a self-storage facility. Besides the obvious sealing of doors and windows, its appearance has been altered by the removal of the ornate (but purely decorative) parapets. H. H. HARWOOD, JR.

the point of hopeless deterioration and was demolished—but then was re-created in its original form.

Those archeologists who also like hiking in the woods can experience what riding the western end of the bucolic Ellicott City car line was like and get their exercise, too. The one-time right of way between Chalfonte Drive and Oella Avenue, including an impressive rock cut at West-chester Avenue, has been converted to a finished trail.

Besides its city streetcar system, Baltimore also had one of the coun-try's finest high-speed electric interurban lines, the Washington, Baltimore & Annapolis. The WB&A died of financial malnutrition in 1935, although its Baltimore-Annapolis segment carried passengers until 1950 under the auspices of the Baltimore & Annapolis Railroad. Despite its durable construction and elaborate engineering, the WB&A left only three relics within Baltimore itself, all of them now so well dis-guised that only the hardest-core railroad historian recognizes them. Yet at least two of them are common sights on well-traveled paths.

The most obscure of the three is the shell of what was once a large, brick electric substation at the northeastern corner of Ostend and Scott Streets, the point where Washington-bound WB&A trains left the city streets and climbed to an impressive steel trestle over the B&O Railroad tracks. Perhaps more familiar is the line's original 1908 Baltimore terminal and general office building, a brick structure that sits in the center of the triangular block bounded by Liberty Street, Park Avenue, and Lexington Street. Later adapted for use as a bank, it gives little indication that it once incorporated an open trainshed on its ground floor. The WB&A vacated the building in 1921, when it moved to a much larger terminal taking up most of the block of Howard-Lombard-Eutaw-Pratt Streets; ironically, the old station far outlived its successor, which was demolished for the downtown Holiday Inn.

Finally, and much more esoteric, is the WB&A's Westport tunnel, which took the high-speed double-track electric line through the hill in the area of Annapolis Road and Waterview Avenue. Although really a rather garden variety concrete subway structure, it was an early example of this type of construction (built in 1907–8) and had the distinction of being one of the very few interurban tunnels in the United States. Baltimore & Annapolis trains used it until February 1950; afterward,

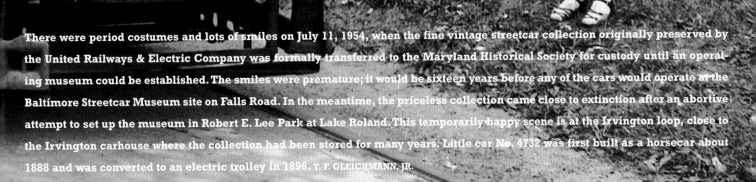

There were period costumes and lots of smiles on July 11, 1954, when the fine vintage streetcar collection originally preserved by the United Railways & Electric Company was formally transferred to the Maryland Historical Society for custody until an operating museum could be established. The smiles were premature; it would be sixteen years before any of the cars would operate at the Baltimore Streetcar Museum site on Falls Road. In the meantime, the priceless collection came close to extinction after an abortive attempt to set up the museum in Robert E. Lee Park at Lake Roland. This temporarily happy scene is at the Irvington loop, close to the Irvington carhouse where the collection had been stored for many years. Little car No. 4732 was first built as a horsecar about 1888 and was converted to an electric trolley in 1896. T. F. GLEICHMANN, JR.

the Baltimore-Washington Parkway was built on this section of the old roadbed. But for some reason the interurban tunnel was bypassed and not touched. Still intact, albeit sealed and refaced with stone, it can be seen along the west side of the parkway just north of the Annapolis Road exit.

South of the city, a lengthy section of the old WB&A right of way not only still exists but is again carrying rail passengers. Beginning at Westport, Baltimore's new light rail line uses portions of the routes of the WB&A and one of its predecessors, the Annapolis Short Line, as far as the present Cromwell terminal in Glen Burnie. And along the route light rail passengers can glimpse the WB&A's former Linthicum station at Church Road. South of the present light rail terminal north of Dorsey Road in Glen Burnie, the old Baltimore & Annapolis right of way is now a well-used hiking-biking trail.

Finally, do not overlook the obvious: the Baltimore Streetcar Museum (BSM). Like the structures just mentioned, the museum's collection covers the full span of Baltimore's transit history, from horsecars to one of the last PCC streamliners. But unlike the one-time shops and car-houses, this slice of history is not static. Part of the historic collection operates regularly and—money, manpower, and natural disasters permitting—all will do so some day. The museum houses a genuinely unique collection—the only one covering a single city's public transportation from end to end; its existence is an amazing combination of foresight (by the old United Railways & Electric Co., which assembled most of it), perseverance (by BSM members, through one near-fatal false start and one flood), and spots of good fortune (with help from Baltimore city government). Baltimore's streetcar era was part of its past life, and no past life really can be brought back. Happily, however, the mementoes are alive.

Today at the Baltimore Streetcar Museum; No. 3828, a carefully restored 1902 Brill product, trundles along the museum's track west of Falls Road. This right of way was itself restored; it was originally the "Ma & Pa" Railroad's main line out of Baltimore. The railroad's one-time freight station is in the right rear and ahead of the car is its stone roundhouse. H. H. HARWOOD, JR.

Printed in China on acid-free paper
First Edition published 1984 by Quadrant Press, Inc., as *Baltimore and Its Streetcars: A Pictorial Review of the Postwar Years*
Revised edition 2003
2 4 6 8 9 7 5 3 1

The Johns Hopkins University Press
2715 North Charles Street
Baltimore, Maryland 21218-4363
www.press.jhu.edu

Jacket and Book Design: Jody Billert and Kristen Fernekes / Design Literate, Inc.

Library of Congress Cataloging-in-Publication Data
Harwood, Herbert H.
[Baltimore and its streetcars]
Baltimore streetcars: the postwar years / Herbert H. Harwood, Jr.; with a foreword by Paul W. Wirtz.
p. cm.
"First edition published in 1984 by Quadrant Press, Inc., as Baltimore and its streetcars: a pictorial review of the postwar years" — T.p verso.
ISBN 0-8018-7190-5
1. Street-railroads—Maryland—Baltimore—History—20th century. I. Title.
HE4491.B372 H37 2003
388.4'6'09752609044—dc21
2002006468

A catalog record for this book is available from the British Library.

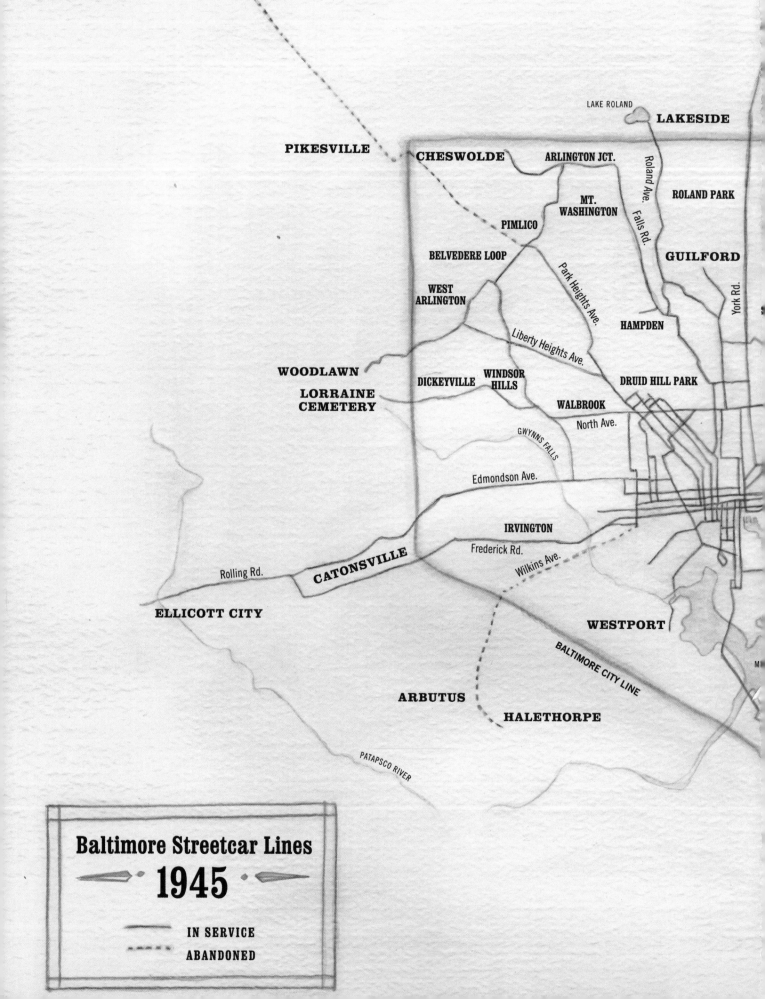

to REISTERSTOWN & GLYNDON

LAKE ROLAND   **LAKESIDE**

PIKESVILLE   **CHESWOLDE**   **ARLINGTON JCT.**

ROLAND PARK

**MT.
WASHINGTON**

PIMLICO

Roland Ave.

Falls Rd.

GUILFORD

BELVEDERE LOOP

**WEST
ARLINGTON**

Park Heights Ave.

York Rd.

Liberty Heights Ave.

HAMPDEN

WOODLAWN

**LORRAINE
CEMETERY**

DICKEYVILLE   **WINDSOR
HILLS**

DRUID HILL PARK

WALBROOK

North Ave.

GWYNNS FALLS

Edmondson Ave.

IRVINGTON

CATONSVILLE

Frederick Rd.

Wilkins Ave.

Rolling Rd.

ELLICOTT CITY

WESTPORT

BALTIMORE CITY LINE

ARBUTUS

HALETHORPE

PATAPSCO RIVER

Baltimore Streetcar Lines
1945

IN SERVICE

ABANDONED